Studies in Computational Intelligence

Volume 602

Series editor

Janusz Kacprzyk, Polish Academy of Sciences, Warsaw, Poland
e-mail: kacprzyk@ibspan.waw.pl

About this Series

The series "Studies in Computational Intelligence" (SCI) publishes new developments and advances in the various areas of computational intelligence—quickly and with a high quality. The intent is to cover the theory, applications, and design methods of computational intelligence, as embedded in the fields of engineering, computer science, physics and life sciences, as well as the methodologies behind them. The series contains monographs, lecture notes and edited volumes in computational intelligence spanning the areas of neural networks, connectionist systems, genetic algorithms, evolutionary computation, artificial intelligence, cellular automata, self-organizing systems, soft computing, fuzzy systems, and hybrid intelligent systems. Of particular value to both the contributors and the readership are the short publication timeframe and the worldwide distribution, which enable both wide and rapid dissemination of research output.

More information about this series at http://www.springer.com/series/7092

Mohamed Medhat Gaber
Mihaela Cocea · Nirmalie Wiratunga
Ayse Goker
Editors

Advances in Social Media Analysis

 Springer

Editors
Mohamed Medhat Gaber
School of Computing Science and Digital
  Media
Robert Gordon University
Aberdeen
UK

Nirmalie Wiratunga
School of Computing Science and Digital
  Media
Robert Gordon University
Aberdeen
UK

Mihaela Cocea
School of Computing
University of Portsmouth
Portsmouth
UK

Ayse Goker
School of Computing Science and Digital
  Media
Robert Gordon University
Aberdeen
UK

ISSN 1860-949X   ISSN 1860-9503   (electronic)
Studies in Computational Intelligence
ISBN 978-3-319-35618-1   ISBN 978-3-319-18458-6   (eBook)
DOI 10.1007/978-3-319-18458-6

Springer Cham Heidelberg New York Dordrecht London
© Springer International Publishing Switzerland 2015
Softcover reprint of the hardcover 1st edition 2015

Printed on acid-free paper

Springer International Publishing AG Switzerland is part of Springer Science+Business Media
(www.springer.com)

Preface

We are happy to present these carefully selected research projects in the area of social media analysis that are organised into seven chapters. The chapters are diverse enough to provide the reader with insights into current research directions. However, owing to the importance of sentiment analysis in social media, there are six chapters that provide the readers with different techniques in this continuously growing area. The other chapter provides an important research direction on how to detect newsworthy topics from social media websites.

Erik Tromp and Mykola Pechenizkiy in the chapter "Pattern-Based Emotion Classification on Social Media" adopt Plutchiks wheel of emotions model and their long-standing rule-based emotion detection method to classify a variety of emotions on social media. Carlos Martin, David Corney and Ayse Goker in the chapter "Mining Newsworthy Topics from Social Media" provide the reader with a number of information retrieval and data mining techniques that are able to identify newsworthy contents in social media websites. Gizem Gezici, Berrin Yanikoglu, Dilek Tapucu and Yücel Saygın in the chapter "Sentiment Analysis Using Domain-Adaptation and Sentence-Based Analysis" motivate sentence-based sentiment analysis as opposed to the lexicon-based approach adopted in a large number of sentiment analysis techniques. In the chapter "Entity-Based Opinion Mining from Text and Multimedia", Diana Maynard and Jonathan Hare prove empirically how multimedia can help resolve the ambiguity of opinion. Such multimodal approach has growing interest with all major social media websites providing means of using multimedia in the users posts. Aminu Muhammad, Nirmalie Wiratunga and Robert Lothian in the chapter "Context-Aware Sentiment Analysis of Social Media" argue that local and global contexts can enhance the performance of sentiment analysis, which has been experimentally proven. In the chapter "Case-Studies in Mining User-Generated Reviews for Recommendation", Ruihai Dong, Michael P. O'Mahony, Kevin McCarthy and Barry Smyth combine topic detection and sentiment analysis for filtering useful reviews and product recommendation. Zheng Yuan and Matthew Purver in the chapter "Predicting Emotion Labels for Chinese Microblog Texts" provide experimental work on predicting emotion in a Chinese microblogging website, namely Sina Weibo using n-gram features, of which higher orders proved to be useful in enhancing the prediction of the emotion.

This volume can serve the audience from both academia and industry, looking for new advances in the area of social media analysis. We hope that the presented chapters open up opportunities for future research.

February 2015   Mohamed Medhat Gaber

Contents

Pattern-Based Emotion Classification on Social Media

Erik Tromp and Mykola Pechenizkiy

Abstract Sentiment analysis can go beyond the typical granularity of polarity that assumes each text to be positive, negative or neural. Indeed, human emotions are much more diverse, and it is interesting to study how to define a more complete set of emotions and how to deduce these emotions from human-written messages. In this book chapter we argue that using Plutchik's wheel of emotions model and a rule-based approach for emotion detection in text makes it a good framework for emotion classification on social media. We provide a detailed description of how to define rule-based patterns for Plutchik's wheel emotion detection, how to learn them from the annotated social media and how to apply them for classifying emotions in the previously unseen texts. The results of the experimental study suggest that the described framework is promising and that it advances the current state-of-the-art in emotion detection.

1 Introduction

Sentiment analysis can be performed at different levels of granularity; the document level [17, 26], word level [12] or the sentence or phrase level [24], and with different levels of detail; determining the *polarity* of a message or the *emotion* expressed [22]. When sentiment analysis is performed on social media, in which a single message

E. Tromp (✉)
Adversitement B.V., Uden, The Netherlands
e-mail: e.t.tromp@gmail.com

M. Pechenizkiy
Eindhoven University of Technology, Eindhoven, The Netherlands
e-mail: m.pechenizkiy@tue.nl
http://www.win.tue.nl/~mpechen/

E. Tromp
Department of Computer Science, TU Eindhoven,
P.O. Box 513, 5600 MB Eindhoven, The Netherlands

© Springer International Publishing Switzerland 2015   1
M.M. Gaber et al. (eds.), *Advances in Social Media Analysis*,
Studies in Computational Intelligence 602,
DOI 10.1007/978-3-319-18458-6_1

typically consists of one or two sentences, we study how sentiment is expressed at the sentence level.

Current sentiment analysis methods—ranging from baseline bag-of-words methods to state-of-the-art recursive neural networks [28]—typically focus on deducing information on subjectivity or polarity only (Sect. 4). Human emotions move far beyond these simple metrics and are much more diverse. This implies that such subjectivity- or polarity-analysis only gives limited information on the actual intent of an author of a message.

Defining axes of polarity is not a hard task, typically one has negativity, positivity and a notion of neutrality or objectivity in between. For emotions however, defining a complete and clear set of emotions is much more difficult. Though several researchers attempted at defining standards in this field [20, 21, 25], AAAC,[1] there is still no consensus on a basic set of emotions that is generally accepted and could be objectively verified.

The goal of this chapter is to present a sentiment analysis approach accompanied by a model of emotions that fit well together in order to set a standard in emotion analysis to expand upon.

To achieve this goal we do not seek to define or implement our own model of emotions, but choose an existing psychological model of basic emotions that is manageable yet applicable to any given domain or language. We provide motivation why the wheel of emotions defined by Plutchik [21] is suitable for our purpose. Besides the model of emotions, we aim to define an algorithm that allows to deduce these emotions from human-written texts.

We present a new *RBEM-Emo* approach [31] for emotion detection from human-written texts. This algorithm is based on work by [30] where the Rule-Based Emission Model (RBEM) algorithm for polarity detection only was introduced. RBEM generates positive and negative emissions based on several groups of patterns that capture various ways how sentiment can be expressed in natural language.

In [30] we extensively experimented with RBEM on English and Dutch messages extracted from Twitter. The experiments demonstrate that designing such an algorithm instead of applying the state-of-the art general purpose classification techniques is a reasonable choice for the automated sentiment classification in practice. Using RBEM we were able to design a competitive sentiment classification system showing promising accuracy results close to 80 % on the considered datasets. We also illustrated that RBEM can be used in multilingual settings and is applicable to social media characterized by use of not always regular language constructs. Besides, we provided some further evidence that RBEM-based systems are easy to debug, improve over time and adapt to new application domains, for which no previously annotated data were available. This is rather important in practice too as use of language is highly dependent upon the domain in which it is being used. As such, it is expected that a generically trained model does not perform as well as it should on a specific domain and that domain-specific models do not port well to other domains. RBEM in fierce contrast to general-purpose state-of-the-art classification techniques

---

[1]The Association for the Advancement of Affective Computing—http://emotion-research.net/.

used for sentiment classification, including e.g. SVMs, supervised sequence embedding [3] or deep learning neural networks [11] with which adaptation of models is a nontrivial labor-intensive process requiring a deep understanding of machine learning, requires little effort of domain expert rather than intense efforts of a machine learning expert.

Such promising results encouraged us to look into ways to use a similar approach for finer-grained sentiment analysis at the level of human emotions. Luckily, extending RBEM to emotion detection appears to be rather straightforward as we present in this chapter. We show how RBEM can be developed further to go beyond polarity and measure emotions as given by Plutchik's wheel of emotions [21].

We introduced RBEM-Emo in preliminary technical report [31]. Here, for the sake of completeness we present both RBEM and RBEM-Emo pattern groups, explain how to induce them from the labeled data and how to use them for emotion classification.

We conducted an experimental evaluation of RBEM-Emo on a publicly available benchmark and on a new benchmark that we constructed. The results of our evaluation suggest that RBEM-Emo outperforms the current state-of-the-art approaches for emotion detection. To facilitate reproducibility of the results and further progress in emotion classification from social media we made our benchmark publicly available.

The rest of the chapter is organized as follows. In Sect. 2 we describe our approach for emotion detection, and Sect. 3 presents its experimental evaluation. We discuss related work in Sect. 4. Section 5 concludes with the lessons learnt and directions for further research.

2 Emotion Detection Framework

The approach we take to emotion detection consists of combining Plutchik's wheel of emotions with the RBEM-Emo algorithm as an emotion classifier on text messages. We next discuss the model of emotions and RBEM-Emo patterns, their construction and use for emotion detection.

2.1 Plutchik's Wheel of Emotions

To tackle the problem of emotion detection, one needs to have a notion of emotion. As e.g. in text mining the problem can be formulated differently depending on whether we have just two classes like in spam filtering, or several categories like topic classification or a large number of categories like in automated tagging. We choose the wheel of emotions defined by Plutchik [21] because it defines only eight basic emotions, which makes the problem manageable for envisioned applications, and because it makes a good match with the proposed RBEM-Emo approach as we detail below.

These eight emotions are assumed to be complete in the sense that any expressed emotion is related or subsumed by one of the eight. In his work, Plutchik states that these emotions are culturally independent. Given this assumption, we can apply this model to any given language, which we consider to be a strong point.

Another reason for using this model is that each of these eight basic emotions are opposites of one of the other basic emotions. This means that we can in fact measure four axes where opposite emotions exist on the two extremes of a single axis. Additionally, Plutchik defines eight *human feelings* that are derivatives of combinations of two basic emotions. This in fact means that with modeling only four axes, we can get a total of sixteen dimensions of emotions and feelings.

Plutchik originally chose to represent his model as a *wheel* of emotions as presented in Fig. 1 where adjacent emotions are related. Moving from inner circles to outer circles should be interpreted as gradually increasing complexity of an emotion. The second most inner circle contains the basic emotions that Plutchik found to be the most complete set of emotions that are culturally independent. Outside of the wheel, derived human feelings are shown. These feelings are composed of the two basic emotions they are adjacent to.

Fig. 1 Plutchik's wheel of emotions [21]

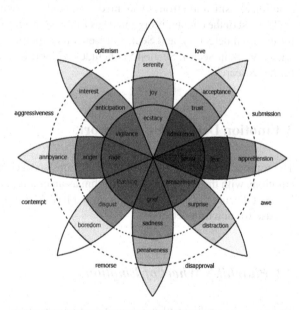

Table 1 The basic emotions and their opposites of Plutchik's model

Basic emotion	Opposite
Joy	Sadness
Trust	Disgust
Fear	Anger
Surprise	Anticipation

Table 2 Human feelings that are composites of two basic emotions

Human feeling	Emotions	Opposite
Optimism	Anticipation, joy	Disapproval
Love	Joy, trust	Remorse
Submission	Trust, fear	Contempt
Awe	Fear, surprise	Aggression
Disapproval	Surprise, sadness	Optimism
Remorse	Sadness, disgust	Love
Contempt	Disgust, anger	Submission
Aggression	Anger, anticipation	Awe

Tables 1 and 2 present a structured overview of all basic emotions and derived human feelings present in Plutchik's model as well as defining their exact opposites. Note that since the basic emotions all have an opposite, the derived, composite human feelings also all have opposites.

2.2 Rule-Based Emotion Classification

The rules used in the RBEM algorithm directly stem from nine different *pattern groups*, defined as follows.[2]

- **Positive** patterns are positive when taken out of context. English examples hereof are *good, well done*.
- **Negative** patterns are negative when taken out of context, e.g. *bad, terrible*.
- **Amplifier** patterns strengthen polarity of n entities to their left and right, either positive or negative, e.g. *very much, a lot*.
- **Attenuator** patterns weaken polarity of n entities to their left and right, either positive or negative, e.g. *a little, a tiny bit*.
- **Right Flip** patterns flip the polarity of n entities to their right, e.g. *not, no*.
- **Left Flip** patterns flip the polarity of n entities to their left, e.g. *but, however*.
- **Continuator** patterns continue the emission of polarity, e.g. *and, and also*.
- **Stop** patterns interrupt the emission of polarity. Stop patterns usually are punctuation signs such as a dot or an exclamation mark, expressing the general case that polarity does not cross sentence boundaries.
- **Neutral** patterns do not have any particular meaning but may eliminate the existence of other patterns in a given context.

The need for positive, negative and negation patterns is evident. The need for continuators and left flips has been indicated in [12]: conjunctive words such as *and* usually

---

[2]Note that the examples only list words but a pattern can consist of any combination of words and POS-tags. This concept is further explained when we describe how to learn a model.

connect adjectives of the same polarity whereas conjunctive words such as *but* usually connect words of opposing polarity. It is easily seen that certain words strengthen or weaken polarity, these are covered by the amplifier and attenuator patterns. The stop patterns are especially useful in determining sentence-based sentiment as these patterns block polarity emission and typically consist of sentence delimiters such as punctuation. The neutral pattern group does not have a specific logic or rule associated with it but is merely there to eliminate the presence of other patterns when a neutral pattern subsumes a pattern of a different pattern group.

Combining these nine pattern groups using some simple rules allows us to define an emissive model.

RBEM uses a pattern matching on wildcards to identify patterns in a message. When classifying previously unseen messages, two steps are performed. First all patterns in the model that match a message are collected. Then, rule(s) associated with each pattern group for each pattern present in the message are applied.

We next describe in detail how a model is constructed and how to classify previously unseen data.

2.3 Learning RBEM

Each message m of length n is represented as a list $m = [(w_1, t_1), \ldots, (w_n, t_n)]$ of tuples of a word w_i with its respective POS-tag t_i. Upon such a message, patterns can be defined. A pattern is a list of tuples of words and POS-tags represented as m. Patterns belong to a certain pattern group and hence we represent a pattern q as a tuple $q = (g, p)$, where g is the pattern group q belongs to, and p is the list of entities comprising the actual pattern. In general, each element (w_i', t_i') of a pattern p consists of a word w_i' which is precisely defined and a POS-tag t_i' which is also precisely defined. As an exception, elements of p may contain wildcards instead. We consider three types of wildcards.

- **Word wildcards** $(\_, t_i')$: in this case we only consider t_i'. w_i' can be any arbitrary word.
- **Single-position wildcards** $(\_, \_)$: in this case a single entity can be any arbitrary combination of a single word and a single POS-tag.
- **Multi-position wildcards** $(*, *)$: in this case any arbitrary combination of word and POS-tag pairs of any arbitrary length matches the pattern.

Note that word and single-position wildcards can occur at any position in p. But multi-position wildcards can only occur in between two elements that are not multi-position wildcards as co-occurrence of other multi-position wildcards yields another multi-position wildcard.

Our model now simply consists of a set of patterns per pattern group, represented as the set *Model*, containing tuples of groups and patterns; (g, p). All patterns except for the positive and negative patterns adhere to an action radius $\mathscr{E}$. We set $\mathscr{E} = 4$ according to the related experimental results with negation patterns reported in [32].

In general it is possible that the optimal choice of $\mathscr{E}$ may vary from pattern to pattern and/or from one language to the other.

2.4 Classifying with RBEM

When classifying previously unseen data, we perform two steps. First we collect all patterns in our model that match our sentence. Then, we apply a rule associated with each pattern group—with exception of the neutral group—for each pattern present in our message.

Pattern Matching. Each pattern $q = (g, p) \in Model$ is matched against our message $h = [(w_1, t_1), \ldots, (w_n, t_n)]$ where $p = [(v_1, s_1), \ldots, (v_m, s_m)]$. We consider each tuple (w_i, t_i) and evaluate $(v_1, s_1) =_{match} (w_i, t_i)$ where $=_{match}$ is defined as follows:

$$(v_j, s_j) =_{match} (w_i, t_i) \equiv$$

$$
\begin{cases}
\text{true} & (1) \\
\quad \text{if } j > m, \text{define } end \leftarrow i \\
\text{false} & (2) \\
\quad \text{if } i > n \\
v_j = w_i \land s_j = t_i \land (v_{j+1}, s_{j+1}) =_{match} (w_{i+1}, t_{i+1}) & (3) \\
\quad \text{if } v_i \neq \_ \land v_i \neq * \land j \leq m \land j \leq n \\
s_j = t_i \land (v_{j+1}, s_{j+1}) =_{match} (w_{i+1}, t_{i+1}) & (4) \\
\quad \text{if } v_i = \_ \land s_i \neq \_ \land j \leq m \land j \leq n \\
(v_{j+1}, s_{j+1}) =_{match} (w_{i+1}, t_{i+1}) & (5) \\
\quad \text{if } v_i = \_ \land s_i = \_ \land j \leq m \land j \leq n \\
(v_{j+1}, s_{j+1}) =_{match} (w_{i+1}, t_{i+1}) \lor (v_j, s_j) = \\
\qquad =_{match} (w_{i+1}, t_{i+1}) & (6) \\
\quad \text{if } v_i = * \land j \leq m \land j \leq n
\end{cases}
$$

Note that in the definition of $=_{match}$, cases (4)–(6) correspond to the three different types of wildcards. Moreover, in the evaluation of the first disjunction of (6), $(v_{j+1}, s_{j+1}) =_{match} (w_{i+1}, t_{i+1})$, it must hold that $v_{j+1} \neq * \land s_{j+1} \neq *$ due to the restriction we put on the occurrence of multi-position wildcards.

We match all patterns of all groups against every possible element (w_i, t_i) of m. While doing this, we need to keep track of two positions if a pattern matches; the start position of the match in m and the end position of the match in m. The starting position is i whereas the end position is end which is assigned a value in case (1) of $=_{match}$, implying a match between the pattern and the message. We thus get a set of matching patterns containing a start position, an end position and a pattern.

$$matchedPatterns = \{(start, end, (g, [(v_1, s_1), .., (v_n, s_n)])) \mid$$
$$(v_1, s_1) =_{match} (w_{start}, t_{start})\}$$

Elements of *matchedPatterns* may subsume each other. Subsumption in this sense is defined as follows, where we say that q_1 subsumes q_2 in message m.

$$\exists_{(s_1,e_1,q_1),(s_2,e_2,q_2)\in matchedPatterns} : s_1 \leq s_2 \wedge e_1 \geq e_2$$
$$\wedge\neg(s_1 = s_2 \wedge e_1 = e_2) \wedge q_1 \neq q_2$$

All patterns that are subsumed by some other pattern are removed. Note that coinciding patterns, having the same start position as well as the same end position, are not removed but as we deal with sets, such coinciding patterns must be of different pattern groups. Also note that it may be that a pattern containing a wild card may match our sentence multiple times from the same starting position. As the definition of $=_{match}$ dictates, we only find and hence maintain the shortest of such matchings. After removing subsumed patterns, the resulting set *maxPatterns* only contains maximal patterns and is defined as follows. Note that this is where the neutral pattern group plays a role. Whenever a neutral pattern exists in a context that subsumes any other pattern, the neutral pattern is kept whereas the other pattern is discarded. During the application of rules however, nothing is done with this neutral pattern, explaining the name of this pattern group.

$$maxPatterns = \{(s, e, q)|(s, e, q) \in matchedPatterns \wedge$$
$$\neg(\exists_{(s',e',q')\in matchedPatterns} : s \leq s' \wedge e' \geq e$$
$$\wedge\neg(s = s' \wedge e = e') \wedge q \neq q')\}$$

Rule Application. After having collected all maximal patterns, we can apply the heuristic rules for each different pattern group, excluding the neutral pattern group. The rules formally work out the motivation for the presence of each pattern group. The order in which the rules are applied is crucial and so is the role of the action radius $\mathscr{E}$. We outline each of the rules in the order in which they are to be applied. We assume we are given a message m and a model (*Model*, $\mathscr{E}$) on which *maxPatterns* is defined. Every element $e_i = (w_i, t_i) \in m$ has a certain emission value $em(e_i)$ which initially is set to 0 for all $e_i \in m$.

Rule 1. Setting Stops—This rule sets emission boundaries in our message m. It uses all left flip and stop patterns and sets a stop at the starting position of such a pattern. We thus get a set of stops:

$$stops = \{s|(s, f, leftflip) \in maxPatterns$$
$$\vee(s, f, stop) \in maxPatterns\}$$

Rule 2. Removing Stops—Stops set in the previous step can be removed by continuator patterns. This however, only happens to the left of a continuator pattern. We thus remove all stops that occur closest to the left of a continuator pattern, taking $\mathscr{E}$ into account:

$$stops = stops \setminus \{t|t \in stops \wedge$$
$$(\exists_{(s,f,continuator)\in maxPatterns} : t \leq s \wedge s - t < \mathscr{E}$$
$$\wedge\neg(\exists_{t'\in stops} : t < t' \leq s))\}$$

Rule 3. Positive Sentiment Emission—A positive pattern can emit positive sentiment among elements of m. The strength of the emission decays over distance and hence we need a decaying function. We use e^{-x} as decaying function, where x is the distance between the positive pattern and an element of m. The choice of the formula e^{-x} is just a choice made by the authors and is not proven to be the optimal formula. As center for the emission, we take the floor of the center of the pattern in m, computed by taking the center of start and end position. We also need to take all stops into account. For each positive pattern, we update the emission values $em(e_i)$ as follows:

$$\forall_{(s,f,positive)\in maxPatterns} : c = \lfloor \frac{s+f}{2} \rfloor \wedge$$
$$(\forall_{e_i \in m} : \neg(\exists_{t \in stops} : c \geq i \Rightarrow i \leq t \leq c \vee i \geq c$$
$$\Rightarrow c \leq t \leq i) \Leftrightarrow em(e_i) = em(e_i) + e^{-i})$$

Rule 4. Negative Sentiment Emission—Negative patterns are dealt with in the same way positive patterns are. The only difference is that our decaying function is now negative, yielding $-e^{-x}$. The updating of emission values happens in the same manner:

$$\forall_{(s,f,negative)\in maxPatterns} : c = \lfloor \frac{s+f}{2} \rfloor \wedge$$
$$(\forall_{e_i \in m} : \neg(\exists_{t \in stops} : c \geq i \Rightarrow i \leq t \leq c \vee i \geq c$$
$$\Rightarrow c \leq t \leq i) \Leftrightarrow em(e_i) = em(e_i) + -e^{-i})$$

Rule 5. Amplifying Sentiment—Amplifier patterns amplify sentiment emitted either by positive or negative patterns. Similar to the decaying function used for positive and negative patterns, amplification diminishes over distance. Moreover, since entities may already emit sentiment, we use a multiplicative function instead of an additive function. The function we use is $1 + e^{-x}$ where x is the distance. Again this formula is just chosen by the authors and not proven to be optimal. In contrast to positive and negative patterns, amplifiers adhere to the action radius $\mathscr{E}$. The emission values are updated as follows:

$$\forall_{(s,f,amplifier)\in maxPatterns} : c = \lfloor \frac{s+f}{2} \rfloor \wedge$$
$$(\forall_{e_i \in m} : (\neg(\exists_{t \in stops} : c \geq i \Rightarrow i \leq t \leq c \vee i \geq c \Rightarrow$$
$$c \leq t \leq i) \wedge 0 < |c - i| < \mathscr{E}) \Leftrightarrow$$
$$em(e_i) = em(e_i) \cdot (1 + e^{-i}))$$

Note the $0 < |c - i| < \mathscr{E}$ clause. This constraint dictates that $|c - i|$ is at least 1 in $1 - e^{-|c-i|}$ (which is our $1 + e^{-x}$ function), thus avoiding the case that we multiply by 0 (when we allow $|c - i| = 0$, we get $1 - e^0 = 0$) and hence completely remove emission values.

Rule 6. Attenuating Sentiment—Attenuator patterns perform the reverse of amplifier patterns and weaken sentiment. To do so, instead of using $1 + e^{-x}$, we use $1 - e^{-x}$:

$$\forall_{(s,f,amplifier)\in maxPatterns} : c = \lfloor \frac{s+f}{2} \rfloor \wedge$$

$$(\forall_{e_i \in m} : (\neg(\exists_{t \in stops} : c \geq i \Leftrightarrow i \leq t \leq c \vee i \geq c$$

$$\Leftrightarrow c \leq t \leq i) \wedge 0 < |c - i| < \mathscr{E}) \Leftrightarrow$$

$$em(e_i) = em(e_i) \cdot (1 - e^{-i}))$$

Rule 7. Right Flipping Sentiment—Right flip patterns simply flip the emission of sentiment to their right as follows. If there is a stop at the exact center of our right flip, we disregard it:

$$\forall_{(s,f,rightflip)\in maxPatterns} : c = \lfloor \frac{s+f}{2} \rfloor \wedge (\forall_{e_i \in m} : (\neg(\exists_{t \in stops} :$$

$$c < t \leq i) \wedge |c - i| < \mathscr{E}) \Leftrightarrow em(e_i) = -em(e_i))$$

Rule 8. Left Flipping Sentiment—Left flip patterns mirror the effect of right flip patterns:

$$\forall_{(s,f,leftflip)\in maxPatterns} : c = \lfloor \frac{s+f}{2} \rfloor \wedge (\forall_{e_i \in m} : (\neg(\exists_{t \in stops} :$$

$$i \leq t < c) \wedge |c - i| < \mathscr{E}) \Leftrightarrow em(e_i) = -em(e_i))$$

Once the above rules have been applied in the order given, every element e_i of m has an emission value $em(e_i)$. The final polarity of the message is defined by the sum of all emission values for all elements of m:

$$polarity = \sum_{i=1}^{n} em(e_i)$$

Straightforwardly, we say that m is *positive* (class $+$) if and only if *polarity* > 0. Likewise, we say that m is *negative* (class $-$) if and only if *polarity* < 0. Whenever *polarity* $= 0$, we say that m is *neutral* (class $=$).

When looking at the rules, it becomes clear that the order is important. Stops need to be set first since the other rules depend on stops. Next positive and negative sentiment need to be defined because amplifying, attenuating and flipping sentiment requires sentiment beforehand. Next the sentiment is amplified and attenuated based on the positive and negative emissions defined before. Finally the flips change the direction of the sentiment.

2.5 Patterns Specific to the Emission of Emotions

The pattern groups described above can be used for polarity detection assigning new messages a label that is one of *positive, neutral, negative*. The algorithm's internals work in such a way that either positive or negative emissions can be generated upon which subsequently different rules are executed to modify these emissions.

Crucial to the algorithm is that positivity and negativity are opposites of each other and hence allow for example negations to simply invert the emission. This specific characteristic of the algorithm makes it work well with Plutchik's model since the emotions defined in that model are also opposites of each other. We in fact extend the RBEM algorithm to perform the same type of rules but now—instead of having one axis to measure; positive on one end of the extreme and negative on the other extreme—we have four different axes, together yielding eight different emotions being measured.

RBEM-Emo extends RBEM for emotion detection by introducing new pattern groups. The RBEM algorithm uses two base pattern groups to define emission of polarity, positive and negative patterns. For our RBEM-Emo algorithm, we replace these two pattern groups with eight new pattern groups, one for each basic emotion of Plutchik's model: *joy, sadness, trust, disgust, fear, anger, surprise, anticipation*. Similarly, we replace the two rules that are defined on positive and negative patterns with eight new rules. Note that conceptually, we perform the exact same process we do for positive polarity on one hand and negative polarity on the other hand, but now four times, once for each axis.

Since we no longer operate on a single emission score but instead on four, we define a mapping from emotions to an index by esc_{emo} and we define a sign counterparts $sign_{emo}$ for each emotion on a single axis. Here $esc_{Joy} = esc_{Sadness} = 1$ and $sign_{Joy} = 1, sign_{Sadness} = -1, esc_{Trust} = esc_{Disgust} = 2$ and $sign_{Trust} = 1, sign_{Disgust} = -1, esc_{Fear} = esc_{Anger} = 3$ and $sign_{Fear} = 1, sign_{Anger} = -1, esc_{Surprise} = esc_{Anticipation} = 4$ and $sign_{Surprise} = 1, sign_{Anticipation} = -1$. We also define a subscripted emission score $em_j(e_i)$ where $j \in [1, 4]$ and the value of j corresponds with the emotion axis for the emotions that map to j using esc_{emo} (i.e. em_1 is the axis function used by *Joy* and *Sadness*).

The new rules that replace the original rules defining positive sentiment emission and negative sentiment emission are defined as follows:

$$\forall_{emo \in \{Joy, Sadness, Trust, Disgust, Fear, Anger, Surprise, Anticipation\}} :$$

$$\forall_{(s, f, emo) \in maxPatterns} : c = \lfloor \frac{s + f}{2} \rfloor \wedge$$

$$(\forall_{e_i \in m} : \neg(\exists_{t \in stops} : c \geq i \Rightarrow i \leq t \leq c \vee i \geq c \Rightarrow c \leq t \leq i)$$

$$\Leftrightarrow em_{esc_{emo}}(e_i) = em_{esc_{emo}}(e_i) + sign_{emo} \cdot e^{-i}).$$

These new rules also replace rule 3 and 4 with respect to the ordering. All the the other original RBEM rules are executed four times, once for every em_j, $j \in [1, 4]$.

When the algorithm terminates, this yields us four emission scores, i.e. one score per dimension.

Once the algorithm has terminated, we can obtain a total score for each pair or opposite emotions, e.g. for Joy and Sadness by summing of all emissions of em_j. $JoySadness = \sum_{i=1}^{n} em_1(e_i)$. Whenever $JoySadness > 0$ we say that Joy was expressed in the original message. Similarly, when $JoySadness < 0$, we say that Sadness was expressed. If $JoySadness = 0$, neither Joy nor Sadness was expressed. The other three emission axes can be interpreted similarly.

As an illustrative example, consider the sentence *I thought I would like the new XYZ phone, but now that I have it, it is a huge disappointment, it makes me angry*. Suppose also that we have the following patterns (Part-of-Speech tags left out for simplicity): $(I * like, Anticipation)$, $(but, Leftflip)$, $(huge, Amplifier)$, $(disappointment, Sadness)$, $(angry, Anger)$. The algorithm would first assign the emotion scores to all parts of the sentence where patterns are found. This would yield the first part emitting negatively on em_4, the third phrase emitting negatively on em_1 and the last phrase emitting negatively on em_3. Next, the scores on pattern indicated by the word *huge* will amplify the emissions on all axes, with the biggest effect on em_1. Finally, the leftflip indicated by *but* will convert all negative emissions on its left—influencing em_4 mainly—to its opposite direction, yielding positive emissions on em_4. The final outcome will hence be that—ordered by decreasing strength— *Sadness, Anger* and *Surprise* are present.

3 Experimental Evaluation

With the experimental study we aim to evaluate the proposed RBEM-Emo algorithm, which is tailored towards Plutchik's model of emotions.

3.1 Experiment Setup

We compare our method against a majority class baseline, Support Vector Machines (SVMs), regression and the recursive auto-encoder of [27] with respect to their generalization accuracies. In [27] five-dimensional sentiment model originating from the Experience Project[3] is introduced. It would be reasonable to evaluate on this dataset, but the five labels used to express emotions in that dataset are quite arbitrary and ambiguous,[4] as the authors already indicate. In addition, these labels are produced by users that read an actual confession by a different person and instead of capturing the emotion of the actual message hence capture the emotion triggered with an external reader.

---

[3] See http://www.experienceproject.com.
[4] The labels are *Sorry, Hugs, You Rock, Teehee, I Understand* and *Wow, Just Wow*.

Due to the impracticalities of the Experience Project dataset for our experiments, we instead benchmark on a different, well-accepted dataset introduced in [1]. This dataset is annotated using Ekman's emotions [10] instead of Plutchik's, but since the six basic emotions of Ekman are subsumed by the eight emotions of Plutchik's model, we can use the labels in a straightforward manner, ignoring labels produced by RBEM-Emo that do not exist in Ekman's model and producing the majority class as label in case we find a non-existing emotion. We refer to this dataset as the *Affect Dataset*.

In addition to benchmarking on a well-accepted public dataset, we also introduce our own *Twitter Dataset* that is annotated on Plutchik's emotions.

For the SVM and regression classification we use LibShortText [33]. We experiment using both word counts and TF-IDF scores as features. For the recursive auto-encoder, we use the Java version referenced to by the authors of [27].[5] To ensure we have the right setup of the auto-encoder, we reproduced the polarity detection experiments on the rotten tomatoes dataset as done in [27] and obtained an accuracy of 77.0 %. This is in line with the results presented in [27], illustrating our setup is valid. When we apply our RBEM-Emo classifier, we get four scores for each axis in Plutchik's model, summing up to eight emotions. Finally, we assign a single label corresponding to the highest of all eight emotion scores.

3.2 Datasets Description

The **Affect Dataset** we use is presented in [1] and is publicly available.[6] This dataset consists of snippets of text obtained from books written by three different authors.

For each snippet, every sentence is annotated by two annotators. These annotators provide two different labels each, one for the prevailing emotion found in the sentence and one for the mood found. The available labels are the six basic emotions of Ekman's universal emotions, being *angry, disgusted, fearful, happy, sad, surprised.* In addition, the authors could also indicate neutrality (Table 3).

We use only those messages for which both annotators agree upon emotion and we discard the mood label produced by the annotators. Moreover, since 85 % of all sentences in the dataset are neutral, and many general purpose classification techniques suffer from class imbalance, we produce two different datasets, one where neutral sentences are removed and only emotion-bearing sentences are maintained and one where neutral messages are included. For evaluation purposes, we use roughly $\frac{2}{3}$ of the data for training and $\frac{1}{3}$ for testing. The resulting sizes of the training sets are 7527 and 1084 instances depending on the in- or exclusion of the neutral class, and for test sets—3590 and 488 instances correspondingly (Table 4).

Twitter Dataset. Since the proposed RBEM-Emo method is tightly integrated with Plutchik's wheel of emotions, we evaluate on data annotated on these emotions. We collected a large amount of tweets in three different languages: *English, Dutch*

---

[5]Can be found at https://github.com/sancha/jrae.
[6]http://lrc.cornell.edu/swedish/dataset/affectdata/.

Table 3 The mapping from emotions in Ekman's model to Plutchik's model

Ekman	Plutchik
Angry	Anger
Disgusted	Disgust
Fearful	Fear
Happy	Joy
Sad	Sadness
Surprised	Surprise

Table 4 The Affect dataset size after removing all sentences that both annotators did not agree upon

	Train	Test
With neutral class	7527	3590
Without neutral class	1084	488

Table 5 The Twitter datasets sizes per language

Language	Train	Test
Dutch	289	113
English	235	113
German	225	109

and *German*. We had at least two independent annotators to annotate each of these messages using a dedicated Web-based annotation tool. In case of disagreement, we use the prevailing emotion label given by the annotators as actual label for a message. If there is no agreement on the prevailing emotion label, the message was discarded.

In addition, the annotators were asked to identify patterns in these messages such that we can later on construct the RBEM-Emo model from them.

The data was collected from Twitter where a language detection algorithm [29] was used to filter out those messages that are written in English, Dutch or German as a first step. All messages wrongly identified by language are later on filtered out by the annotators.

In line with the setup of the experiments presented in [27] and adhered to here, we randomly split the data into roughly $\frac{2}{3}$ training and $\frac{1}{3}$ test data. The resulting training/test set sizes are Dutch 289/113 for Dutch, 235/113 for English and 225/109 for German (Table 5).

The Twitter dataset is made publicly available.[7]

3.3 Results

The accuracies of the best performing general purpose classification techniques on the Affect Dataset are compared to those of RBEM-Emo in Table 6. The majority class classification accuracy is given as a baseline. We report accuracies both for the

---

[7]http://www.win.tue.nl/~mpechen/projects/smm/.

case when neutral messages are kept in our dataset and when they are filtered out. We do this since the neutral messages compose 85 % of the entire original dataset and it is expected that generic classification techniques will suffer from class imbalance and learn biases towards this data rather than find actual emotions. This is reflected in the accuracies of the SVM and regression classifiers which are marginally higher than the majority class baseline. Surprisingly, the recursive auto-encoder (RAE) that is currently claimed to be the state-of-the-art technique for emotion classification performs worse than several simpler classifiers and in fact is as good as a majority class classifier. One possible reason for this might be that the size of our dataset is relatively small. RBEM-Emo classifier being a tailor approach to deduce emotional patterns outperforms the other classifiers.

In the second column of Table 6, we report the accuracies when all messages belonging to the neutral class are removed, yielding a more class-balanced dataset. Here we see much better improvements over the majority class baseline for SVM and regression and now also for the recursive auto-encoder. Using TF-IDF scores for features is favored over using just word counts. The RBEM-Emo method however, still outperforms the other classifiers.

Table 7 lists the accuracies obtained per language on our own Twitter corpus. For each classifier, we report the accuracy on each language (being Dutch, English and German) and report a total accuracy which is the average accuracy over all messages in all three languages. A generic result over all classifiers is that the accuracies on English data seem to be the lowest, implying most ambiguity within this language. Remarkable is that the recursive auto-encoder performs worse than SVM and regres-

Table 6 Accuracies on the Affect dataset

Method	Acc. w/Ntl (%)	Acc. no Ntl (%)
Majority	84.4	37.7
SVM, W.C	86.2	61.3
SVM, TF-IDF	86.2	65.0
Regr., W.C	85.8	59.5
Regr., TF-IDF	85.5	63.4
RAE	84.4	60.4
RBEM-Emo	88.4	67.1

Table 7 Accuracies on the Twitter dataset

Language	Majority	SVM W.C.	SVM TF-IDF	Regr W.C.	Regr TF-IDF	RAE	RBEM-Emo
Nl	50.4	53.1	54.9	53.1	53.1	53.1	56.7
En	42.5	46.0	42.5	45.1	42.5	31.0	47.2
De	34.9	46.8	47.7	40.4	46.8	44.0	53.2
All	42.7	48.7	48.4	46.3	47.5	42.7	52.4

sion models and yields no benefit over the majority class guess. Again, this could be due to the small size of the corpus or difficulty in finding the most suitable model parameters. There is no clear evidence on whether TF-IDF scores or word counts work better for this dataset. The RBEM-Emo classifiers yields the highest accuracy for each of three languages.

4 Related Work

Moving beyond polarity in sentiment analysis is currently upcoming and not well studied yet. Few examples can be found where novel methods are introduced to capture more information than just polarity such as the work of [27] where a recursive auto-encoder is used to predict sentiment distributions in five dimensions. The works of [5, 6] promote affective computing using a framework they call SenticNet. The sentiment dimensions of this framework are modeled in an hourglass-model which is a derivative of Plutchik's wheel of emotions [21]. In [14] the author collected and experimented with a large collection of tweets with self-labeled emotion hashtags.

The closest work to our approach is [2], in which the authors considered a rule-based approach based on a set of positive and negative patterns and valence shifters for handling negations and other linguistic constructs defining the sentiment of a sentence.

Standards on emotion frameworks are difficult to define as emotions are usually subjective and cannot be crisply defined. Works of [20, 21, 25] do aim to define standards in this area by defining a minimal set of basic emotions from which more complex ones can be derived or constructed by combining basic emotions. In [5] the authors develop methods to reason about emotions. In [10], facial expressions are linked to emotions and a final six universal basic emotions are presented.

Polarity detection has been studied in different communities and in different application domains. The polarity of adjectives was studied in [12] with the use of different conjunctive words. A comprehensive overview of the performance of different machine leaning approaches on polarity detection were presented in [17–19]. Typically, polarity detection is solved using supervised learning methods but more recently attention is being paid to unsupervised approaches [16].

Some of the recent works adopt a concept-level approach to sentiment analysis [8], which leverages on common sense knowledge for deconstructing natural language text into sentiments. A notable example is [7], in which a two-level affective common sense reasoning framework is proposed to mimic the integration of conscious and unconscious reasoning for sentiment analysis using data mining techniques.

The idea of using patterns arises from [32] who label subjective expressions (patterns) in their training data. Nevertheless, in their experiments they limit themselves to matching against single-word expressions. The use of rules stems from a different domain. The Brill tagger [4] for POS-tagging uses rules. We borrow this ideology and apply it to polarity and emotion detection. The emission aspect of our algorithm is related to smoothing which is often applied in different machine learning settings. RBEM and RBEM-Emo have also close resemblance to [23] where different rules and patterns are defined on the top of a full linguistic parser output. As our algorithm requires only POS-tags as additional linguistic information, it can cover a broader variety of languages since models for POS-tagging methods are more widely available than linguistic parser models are.

More recently attention is being paid to sentiment analysis on social media [13].

The use of emotion classification in different application is still emerging as automated and accurate approaches for emotion detection from text just started to appear. One of possible applications for emotion classification is churn prediction—the challenging task of determining whether and when a specific currently existing customer or user of a service or product will decide to no longer be a customer, usually due to dissatisfaction. The work of [9] shows that emotions successfully detected from e-mails can help to predict churn. One can imagine how the basic emotion *anticipation* might be a direct pointer of churn. In a similar fashion, the human feeling of *agression*, being a combination of *anger* and *anticipation* might be a good sensor when talking about another application area of emotions; threat detection. In [15] the author provides empirical evidence that emotion analysis can help identify personality.

5 Conclusions

In this work we have described a rule-based classification technique called RBEM-Emo for emotion classification on social media. This emotion classification approach is tightly coupled with the Plutchik's model of emotions. We proposed to use this model because it is relatively compact yet complete and models emotions as opposites of each other, a feature that works well with RBEM-Emo.

The results of our experimental study show that RBEM-Emo is competitive to the current state-of-the-art approaches to sentiment and emotion classification. We experimentally compared the performance of RBEM-Emo against other state-of-the-art general-purpose classification approaches. We did this on two different datasets; one of which was introduced by ourselves and annotated by multiple annotators in the dimensions of Plutchik's model.

In all of our experiments, the RBEM-Emo showed the highest generalization accuracy results. Surprisingly, the recursive auto-encoder—considered to be the state-of-the-art in sentiment analysis, including emotion detection—was outperformed not only by RBEM, but also by simple generic classifiers including SVM and regression and in two out of three cases did not even yield any improvement over the majority class guess. The reason for the poor performance of the auto-encoder is not fully clear, but we think the size of the dataset being relatively small might play a key role here. On the other hand though, the cleaned and filtered Experience Project dataset used in the experiments in [27] consisted of roughly 6000 messages whereas our Affect dataset with neutral messages included roughly 10,000 messages. We also ran the LibShortText classifiers (SVM and regression) on the original Experience Project data and in fact obtained results that are on-par with the best reported accuracy in [27] (we obtained an accuracy of 50.9 % using SVM with word counts, where [27] obtain a highest accuracy of 50.1 %).

New approaches for emotion classification appear every year. It is important to facilitate an easy way to benchmark and compare their performance. For studying emotion classification with Plutchik's model, we developed a new benchmark with carefully annotated Twitter messages in three different languages.

5.1 Future Work

An important question for future work is whether or not the chosen model of emotions is actually expressed well in text. As indicated in the related work, many models of emotions exist and one of such models might be favored over the other, i.e. one model might have a set of emotions that are better expressible in text than the others. We currently have no solid evidence of suitability of one model over the other in this domain which makes it an interesting future study.

The Affect and Twitter datasets contain messages written in one of the three Western-European languages, either Roman or Germanic by nature. It would be interesting to see how RBEM-Emo performs when applied to languages from other groups, for example Slavic languages or Asian languages.

RBEM-Emo is rather straightforward to extend, e.g. adding more patterns and rules that would further increase its accuracy, including domain specific knowledge of the particular source, e.g. Twitter hashtags. Enriching the model via the relevance feedback from the user is also feasible, and automating this process is one of the directions of our further work. We also plan to explore methods to find patterns in an automated fashion rather than through a manual labeling process.

Finally, emotion detection as studied in this work, can be used in many applications. It would be interesting to study actual utility of models of emotions and approaches for emotion classification.

Acknowledgments This research is partly supported by Data Science Center (DSC/e) of TU Eindhoven.

References

1. Alm, E.: Affect in text and speech. Ph.D. thesis (2008)
2. Andreevskaia, A., Bergler, S.: Clac and clac-nb: Knowledge-based and corpus-based approaches to sentiment tagging. In: Proceedings of the 4th International Workshop on Semantic Evaluations, SemEval '07, pp. 117–120, Stroudsburg, PA, USA, Association for Computational Linguistics (2007)
3. Bespalov, D., Qi, Y., Bai, B., Shokoufandeh, A.: Sentiment classification with supervised sequence encoder. In: Proceedings of European Conference on Machine Learning and Principles and Practice of Knowledge Discovery in Databases (ECML-PKDD), vol. LNCS 7523, pp. 159–174. Springer (2012)

4. Brill, E.: A simple rule-based part of speech tagger. In: Proceedings of the Third Conference on Applied natural language processing (ANCL'92), pp. 152–155. Association for Computational Linguistics (1992)
5. Cambria, E., Havasi, C., Hussain, A.: Senticnet 2: a semantic and affective resource for opinion mining and sentiment analysis. In: Proceedings of the Twenty-Fifth International Florida Artificial Intelligence Research Society Conference. AAAI Press (2012)
6. Cambria, E., Hussain, A.: Sentic Computing: Techniques, Tools, and Applications. Springer, Heidelberg (2012)
7. Cambria, E., Olsher, D., Kwok, K.: Sentic activation: a two-level affective common sense reasoning framework. In: Proceedings of AAAI, pp. 186–192 (2012)
8. Cambria, E., Schuller, B., Xia, Y., Havasi, C.: New avenues in opinion mining and sentiment analysis. IEEE Intell. Syst. **28**(2), 15–21 (2013)
9. Coussement, K., Van den Poel, D.: Improving customer attrition prediction by integrating emotions from client/company interaction emails and evaluating multiple classifiers. Expert Syst. Appl. **36**(3), 6127–6134 (2009)
10. Ekman, P.: The argument and evidence about universals in facial expressions of emotion. In: Wagner, H., Manstead, A. (eds.) Handbook of Social Psychophysiology, Wiley Handbooks of Psychophysiology, pp. 143–164. Wiley, Chichester (1989)
11. Glorot, X., Bordes, A., Bengio, Y.: Domain adaptation for large-scale sentiment classification: a deep learning approach. In: Proceedings of the 28th International Conference on Machine Learning, ICML 2011, pp. 513–520 (2011)
12. Hatzivassiloglou, V., McKeown, K.: Predicting the semantic orientation of adjectives. In: Proceedings of the ACL, pp. 174–181 (1997)
13. Kumar, S., Morstatter, F., Liu, H.: Twitter Data Analytics. Springer, New York (2014)
14. Mohammad, S.: #emotional tweets. In: *SEM 2012: The First Joint Conference on Lexical and Computational Semantics—Volume 1: Proceedings of the main conference and the shared task, and Volume 2: Proceedings of the Sixth International Workshop on Semantic Evaluation (SemEval 2012), pp. 246–255, Montréal, Canada, 7–8 June 2012. Association for Computational Linguistics
15. Mohammad, S., Kiritchenko, S.: Using nuances of emotion to identify personality. In: Proceedings of the International Conference on Weblogs and Social Media (ICWSM-13), Boston, MA (2013)
16. Paltoglou, G., Thelwall, M.: Twitter, myspace, digg: unsupervised sentiment analysis in social media, vol. 3, pp. 66:1–66:19, ACM, New York, NY, USA (2012)
17. Pang, B., Lee, L.: A sentimental education: sentiment analysis using subjectivity summarization based on minimum cuts. In: Proceedings of the ACL, pp. 271–278 (2004)
18. Pang, B., Lee, L.: Opinion mining and sentiment analysis. In: Foundations and Trends in Information Retrieval (2008)
19. Pang, B., Lee, L., Vaithyanathan, S.: Thumbs up? sentiment classification using machine learning techniques. In: Proceedings of Conference on Empirical methods in natural language processing (EMNLP'02), pp. 79–86. Association for Computational Linguistics (2002)
20. Parrott, W.: Emotions in Social Psychology. Psychology Press, Philadelphia (2001)
21. Plutchik, R.: A general psychoevolutionary theory of emotion, pp. 3–33. Academic press, New York (1980)
22. Potena, D., Diamantini, C.: Mining opinions on the basis of their affectivity. In: 2010 International Symposium on Collaborative Technologies and Systems (CTS), pp. 245–254 (2010)
23. Remus, R., Hänig, C.: Towards Well-grounded Phrase-level Polarity Analysis, pp. 380–392. Springer, New York (2011)
24. Riloff, E., Wiebe, J., Wilson, T.: Learning subjective nouns using extraction pattern bootstrapping. In: Proceedings of the 7th Conference on Natural Language Learning, pp. 25–32 (2003)
25. Schroder, M., Baggia, P., Burkhardt, F., Pelachaud, C., Peter, C., Zovato, E.: Emotional—an upcoming standard for representing emotions and related states. In: Proceedings of the 4th International Conference on Affective Computing and Intelligent Interaction—Volume Part I, ACII'11, pp. 316–325, Springer, Heidelberg (2011)

26. Sindhwani, V., Melville, P.: Document-word co-regularization for semi-supervised sentiment analysis. In: Eighth IEEE International Conference on Data Mining (ICDM'08), pp. 1025–1030 (2008)
27. Socher, R., Pennington, J., Huang, E., Ng, A., Manning, C.: Semi-supervised recursive autoencoders for predicting sentiment distributions. In: Proceedings of the Conference on Empirical Methods in Natural Language Processing, EMNLP'11, pp. 151–161, Stroudsburg, PA, USA (2011), Association for Computational Linguistics
28. Socher, R., Perelygin, A., Wu, J.Y., Chuang, J., Manning, C.D., Ng, A.Y., Potts, C.: Recursive deep models for semantic compositionality over a sentiment treebank (2013)
29. Tromp, E., Pechenizkiy, M.: Graph-based n-gram language identification on short texts. In: Proceedings of the Twentieth Belgian Dutch Conference on Machine Learning (Benelearn 2011), pp. 27–34 (2011)
30. Tromp, E., Pechenizkiy, M.: RBEM: a rule based approach to polarity detection. In: Proceedings of the Second International Workshop on Issues of Sentiment Discovery and Opinion Mining, WISDOM **2013** (2013)
31. Tromp, E., Pechenizkiy, M.: Rule-based emotion detection on social media: putting tweets on plutchik's wheel. CoRR, abs/1412.4682 (2014)
32. Wilson, T., Wiebe, J., Hoffmann, P.: Recognizing contextual polarity in phrase-level sentiment analysis. In: HLT '05: Proceedings of the conference on Human Language Technology and Empirical Methods in Natural Language Processing, pp. 347–354 (2005)
33. Yu, H., Ho, C., Juan, Y., Lin, C.: Libshorttext: A library for short-text classification and analysis (2013)

Mining Newsworthy Topics from Social Media

Carlos Martin, David Corney and Ayse Goker

Abstract Newsworthy stories are increasingly being shared through social networking platforms such as Twitter and Reddit, and journalists now use them to rapidly discover stories and eye-witness accounts. We present a technique that detects "bursts" of phrases on Twitter that is designed for a real-time topic-detection system. We describe a time-dependent variant of the classic *tf-idf* approach and group together bursty phrases that often appear in the same messages in order to identify emerging topics. We demonstrate our methods by analysing tweets corresponding to events drawn from the worlds of politics and sport, as well as more general mainstream news. We created a user-centred "ground truth" to evaluate our methods, based on mainstream media accounts of the events. This helps ensure our methods remain practical. We compare several clustering and topic ranking methods to discover the characteristics of news-related collections, and show that different strategies are needed to detect emerging topics within them. We show that our methods successfully detect a range of different topics for each event and can retrieve messages (for example, tweets) that represent each topic for the user.

1 Introduction

The growth of social networking sites, such as Twitter, Facebook and Reddit, is well documented. Every day, a huge variety of information on different topics is shared by many people. Given the real-time, global nature of these sites, they are used by many people as a primary source of news content [28]. Increasingly, such sites

C. Martin (✉) · D. Corney · A. Goker
IDEAS Research Institute, School of Computing & Digital Media,
Robert Gordon University, Aberdeen AB10 7QB, Scotland, UK
e-mail: cmdanca@gmail.com

D. Corney
e-mail: dpacorney@gmail.com

A. Goker
e-mail: a.s.goker@rgu.ac.uk

© Springer International Publishing Switzerland 2015
M.M. Gaber et al. (eds.), *Advances in Social Media Analysis*,
Studies in Computational Intelligence 602,
DOI 10.1007/978-3-319-18458-6_2

are also used by journalists, partly to find and track breaking news but also to find user-generated content such as photos and videos, to enhance their stories. These often come from eye-witnesses who would be otherwise difficult to find, especially given the volume of content being shared.

Our overall goal is to produce a practical tool to help journalists and news readers to find newsworthy topics from message streams without being overwhelmed. Note that it is not our intention to re-create Twitter's own "trending topics" functionality. That is usually dominated by very high-level topics and memes, defined by just one or two words or a name and with no emphasis on 'news' nor any attempt to explain *why* something is trending.

The scale and diversity of these sites raise the question: how can users (whether journalists or non-professional "news consumers") find newsworthy topics from sites such as Twitter? One option would be for them to simply identify and follow some Twitter accounts that tend to Tweet regularly about the news, such as @CNN or @BBCNews. This approach has drawbacks however. Major news organizations tend to follow similar agendas, meaning that when an event occurs, either all the accounts will send equivalent messages, flooding the user with redundant messages, or none will send messages and the consumer will never learn of the story. The summer 2013 protests in Gezi Park, Turkey, were largely ignored by the Turkish national media for example and global mainstream media reports initially lagged behind social media reports. Such an approach will also miss what might be termed secondary messages from other sources, such as eye-witnesses, who may provide interesting and informative details about a story.

A similar problem occurs with a second possible option, namely using keywords (including hashtags) to filter incoming messages. This recasts the task as a search task with the attendant risks: all tweets that contain the search terms will be retrieved, including repetitions and redundant messages, while tweets that are relevant but do not contain the specified terms will be missed. A third option is to rely on Twitter's own "trending topics" algorithm, but, as noted above, this makes no attempt to filter for newsworthiness, and so tends to be dominated by celebrity news and Twitter memes.

Our system works by identifying phrases that show a sudden increase in frequency (a "burst") and then finding co-occurring groups of phrases to identify topics. Such bursts are typically responses to real-world events. In this way, the news consumer can avoid being overwhelmed by redundant messages, even if the initial stream is formed of diverse messages. The emphasis is on the temporal nature of message streams as we bring to the surface groups of messages that contain suddenly-popular phrases. An early version of this approach was recently described [2, 23], where it compared favourably to several alternatives and benchmarks. Here we expand and update that work, examining the effect of different clustering and topic ranking approaches used to form coherent topics from bursty phrases.

2 Related Work

Many of the individual techniques we use (as described in Sect. 3) have been used in related work, but not together and not in a user-centred way. No other study that we are aware of focuses on the needs of journalist and news-reading users, and with an emphasis on recency to augment traditional collection statistics such as *tf-idf*. It is our view that the domain must be tackled with a combination of such methods. We build on the idea of the importance of time and the concept of "sessions" now common in query-log analysis [18], but adapt it to the context of emerging news from Twitter.

Newman [27] discusses the central use of social media by news professionals, such as hosting live blogs of ongoing events. He also describes the growth of collaborative, networked journalism, where news professionals draw together a wide range of images, videos and text from social networks and provide a curation service. Broadcasters and newspapers can also use social media to increase brand loyalty across a fragmented media marketplace. Further examples of the use of live blogging by newspapers are given by Thurman et al. [41].

Schifferes et al. [38] discuss many of the issues around building a user-centred tool for professional journalists that identifies and verifies news from online social media. They discuss several examples of how information and misinformation has been spread rapidly through social media, showing both the potential benefits and risks of using Twitter as a news source. They interviewed a number of senior journalists, specialising in social media for mainstream news organizations, who expressed dissatisfaction with the tools currently available. Schifferes et al. suggest that independent measures of the reliability of contributors, content and context can help identify unreliable news and they describe a prototype verification system for automatic topic detection.

Petrovic et al. [32] focus on the task of first-story detection (FSD), which they also call "new event detection". They use a locality sensitive hashing technique on 160 million Twitter posts, hashing incoming tweet vectors into buckets in order to find the nearest neighbour and hence detect new events and track them. This work is extended in Petrovic et al. [33] using paraphrases for first story detection on 50 million tweets. Their FSD evaluation used newswire sources rather than Tweets, based on the existing TDT5 datasets. The Twitter-based evaluation was limited to calculating the average precision of their system, by getting two human annotators to label the output as being about an event or not. This contrasts with our goal here, which is to measure and improve the topic-level recall, i.e. to count how many newsworthy stories the system retrieved.

Benhardus [5] uses standard collection statistics such as *tf-idf*, unigrams and bigrams to detect trending topics. Two data collections are used, one from the Twitter API and the second being the Edinburgh Twitter corpus containing 97 million tweets, which was used as a baseline with some natural language processing used (e.g. detecting prepositions and conjunctions). The research focused on general trending topics (typically finding personalities and for new hashtags) rather than focusing the needs of journalistic users and news readers.

Shamma et al. [39] focus on "peaky topics" (topics that show highly localized, momentary interest) by using unigrams only. The focus of the method is to obtain peak terms for a given time slot when compared to the whole corpus rather than over a given time-frame. The use of the whole corpus favours batch-mode processing and is less suitable for real-time and user-centred analysis.

Phuvipadawat and Murata [34] analysed 154,000 tweets that contained the hashtag "#breakingnews". They determine popularity of messages by counting retweets and detecting popular terms such as nouns and verbs. This work is taken further with a simple *tf-idf* scheme that is used to identify similarity [35]; named entities are then identified using the Stanford Named Entity Recogniser in order to identify communities and similar message groups. Sayyadi et al. [37] also model the community to discover and detect events on the Live Labs SocialStream platform, extracting keywords, noun phrases and named entities. Ozdikis et al. [30] also detect events using hashtags by clustering them and finding semantic similarities between hashtags, the latter being more of a lexicographic method.

Ratkiewitcz et al. [36] focus specifically on the detection of a single type of topic, namely political abuse. Evidence used include the use of hashtags and mentions. Alvanaki [3] propose a system based on popular seed tags (tag pairs) which are then tracked, with any shifts detected and monitored. Becker et al. [4] also consider temporal issues by focusing on the online detection of real world events, distinguishing them from non-events (e.g. conversations between posters). Clustering and classification algorithms are used to achieve this. Methods such as *n*-grams and NLP are not considered. These methods do use natural language processing methods or *n*-grams, but many consider temporal factors in some way.

3 Methods

In this section we describe various aspects of our approach to topic detection and discuss how they work together. We consider "temporal document frequency-inverse document frequency" as a variation of the classic *tf-idf* to find trending terms at a specific point in time. We discuss several clustering methods to group these terms into topic-specific clusters and the use of *n*-grams to find phrases rather than isolated terms. We also consider the optimum speed with which to update results in real time, and compare methods to rank the results most usefully. In our experiments, we use collections of tweets (see Sect. 4.1), but the same approach should work for other streams of text messages.

3.1 BNgrams

Term frequency-inverse document frequency, or *tf-idf*, has been used for indexing documents since it was first introduced [40]. We are not interested in indexing documents however, but in finding novel trends, so we want to find terms that appear in one *time period* more than others. We treat temporal windows as documents and use them to detect words and phrases that are both new and significant. We therefore define newsworthiness as the combination of novelty and significance. We can maximise *significance* by filtering tweets either by keywords (as in this work) or by following a carefully chosen list of users, and maximise *novelty* by finding bursts of suddenly high-frequency words and phrases.

We select terms with a high "temporal document frequency-inverse document frequency", or $df - idf_t$, by comparing the most recent x messages with the previous x messages and count how many contain the term. We regard the most recent x messages as one "slot". After standard tokenization and stop-word removal, we index all the terms from these messages. For each term, we calculate the document frequency for a set of messages using df_{ti}, defined as the number of messages in a slot i that contain the term t.

$$df - idf_{ti} = (df_{ti} + 1) \cdot \frac{1}{\log \left(df_{t(i-1)} + 1 \right) + 1}. \tag{1}$$

This produces a list of terms which can be ranked by their $df - idf_t$ scores. Note that we add one to term counts to avoid problems with dividing by zero or taking the log of zero. To maintain some word order information, we define terms as n-grams, i.e. sequences of n words. Based on experiments reported elsewhere [23], we use 1-, 2- and 3-g in this work. High frequency n-grams are likely to represent semantically coherent phrases. Having found bursts of potentially newsworthy n-grams, we then group together n-grams that tend to appear in the same tweets. Each of these clusters defines a topic as a list of n-grams. We call this process of finding bursty n-grams "BNgrams."

3.2 Topic Clustering

An isolated word or phrase is often not very informative, but a group of them can define the essence of a story. Therefore, we group the most representative n-grams into clusters, each representing a single topic. A group of messages that discuss the same topic will tend to contain at least some of the same n-grams. We can then find the message that contains the most of these n-grams that define a topic, and use that message as the basis of a human-readable label for the topic. We now discuss three clustering algorithms that we compare here.

3.2.1 Hierarchical Clustering

Here, we initially assign every n-gram to its own singleton cluster, then follow a standard "group average" hierarchical clustering algorithm [26] to iteratively find and merge the closest pair of clusters. We define the similarity between two n-grams as the fraction of messages in the same slot that contain both of them, so it is highly likely that the term clusters whose similarities are high represent the same topic.

The clustering is repeated until the similarity between the nearest un-merged clusters falls below a fixed threshold θ, producing the final set of topic clusters for a set of tweets. In our experiments, we use a similarity threshold of $\theta = 0.5$ which means that two terms must appear in at least half of the same tweets in order to belong to the same topic. Note that this threshold implicitly defines the number of clusters that the system returns for any given set of tweets. If we use a very low threshold, then we will merge clusters that only share a few terms, which will tend to lead to a very small number of large clusters. A high threshold will conversely lead to a very large number of small clusters with very few overlapping terms. This gives a potential means to control the granularity of detected topics. Preliminary results suggest that the exact value is not critical.

Individual messages are then assigned to the cluster that they share the most terms with, if any. Note that not every tweet will be assigned to any topic. This is deliberate, as many tweets are not newsworthy and/or do not fall into the same topic category as any other tweets.

Further details about this algorithm and its parameters can be found in our previous published work [2].

3.2.2 Apriori Algorithm

The Apriori algorithm [1] finds all the associations between the most representative n-grams based on the number of tweets in which they co-occur. Each association is a candidate topic at the end of the process. One of the advantages of this approach is that one n-gram can belong to different associations (i.e. it allows partial membership), avoiding one problem with hierarchical clustering. The number of associations does not have to be specified in advance. We also obtain maximal associations after clustering to avoid large overlaps in the final set of topic clusters.

One parameter associated to this technique is the *support value* which determines the minimum number of documents a group of n-grams (association) should share to be considered as a candidate topic. The value of this parameter represents a percentage of all the documents from the corresponding slot. Preliminary experiments considering different values of this parameter suggested we fix its value to 0. It means that no candidate topic is discarded. In addition, maximal associations are obtained at the end of the approach to avoid overlaps in the final candidate topics set. The main idea of this approach is to delete all the associations whose keywords are contained in another association and sharing most of the topic tweets with the previous one. This second requirement was introduced to confirm that both topics are talking about the same matter before they are merged into a single topic.

3.2.3 Gaussian Mixture Models (GMM)

GMMs assign probabilities (or strengths) of membership of each n-gram to each cluster, allowing partial membership of multiple clusters. This approach does require the number of clusters to be specified in advance, although this can be automated (e.g. by using Bayesian information criteria [15]). Here, we use the Expectation—Maximisation algorithm to optimise a Gaussian mixture model [12]. We fix the number of clusters at 20, although initial experiments showed that using more or fewer produced very similar results. A more sophisticated variation would be to vary this value as a function of the number of messages in the slot. Seeking more clusters in the data than there are newsworthy topics means that some clusters will contain irrelevant tweets and outliers, which can later be assigned a low rank and effectively ignored, leaving us with a few highly-ranked clusters that are typically newsworthy.

We use the Weka implementation [17], which iteratively fits spherical Gaussian components to the data.

3.3 Topic Ranking

To maximise usability we need to avoid overwhelming the user with a very large number of topics. We therefore want to rank (and potentially filter) the results by relevance, in the same fashion as typical search engines. Here, we compare two topic ranking techniques.

3.3.1 Maximum n-gram $df - idf_t$

One method is to rank topics according to the maximum $df - idf_t$ value of their constituent n-grams. The motivation of this approach is the assumption that the most popular n-gram from each topic represents the core of the topic.

3.3.2 Weighted Topic-Length

As an alternative we propose weighting the topic-length (i.e. the number of terms found in the topic) by the number of tweets in the topic to produce a score for each topic. Thus the most detailed and popular topics are assigned higher rankings. The use of clustering techniques that allow each n-gram to have partial membership of different clusters suggests the need for an alternative topic ranking technique, because the previous method may fail to give a good performance if the top-m results from the ranking have several and diverse topics at the same time. We define this score thus:

$$s_t = \alpha \cdot \frac{L_t}{L_{max}} + (1 - \alpha) \cdot \frac{N_t}{N_s} \qquad (2)$$

where s_t is the score of topic t, L_t is the length of the topic, L_{max} is the maximum number of terms in any current topic, N_t is the number of tweets in topic t and N_s is the number of tweets in the slot. Finally, α is a weighting term. Setting α to 1 rewards topics with more terms; setting α to 0 rewards topics with more tweets. We used $\alpha = 0.7$ in our experiments, giving slightly more weight to those stories containing more details, although the exact value is not critical.

4 Experiments

Here, we show the results of our experiments with several variations of the BNgram approach. We focus on two questions. First, what is best slot size to balance topic recall and refresh rate? A very small slot size might lead to missed stories as too few tweets would be analysed; conversely, a very large slot size means that topics would only be discovered some time after they have happened. This low 'refresh rate' would reduce the timeliness of the results. Second, what is the best combination of clustering and topic ranking techniques? In Sect. 3, we introduced three clustering methods and two topic ranking methods; we need to determine which methods are most useful.

We have previously shown that our methods perform well [2]. The BNgram approach was compared to a popular baseline system in topic detection and tracking—Latent Dirichlet Allocation (LDA) [6]—and to several other competitive topic detection techniques, getting the best overall topic recall. In addition, we have shown the benefits of using n-grams compared with single words for this sort of analysis [23]. Below, we present and discuss the results from our current experiments, starting with our approach to evaluation.

4.1 Evaluation Methods

When evaluating any information retrieval system, it is crucial to define a realistic test problem. We used three Twitter data sets focused on popular real-world events and compare the topics that our algorithm finds with an externally-defined ground truth. To establish this ground truth, we relied on mainstream media (MSM) reports of the three events. This use of MSM sources helps to ensure that our ground truth topics are newsworthy (by definition) and that the evaluation is goal-focussed (i.e. will help journalists write such stories). We see no reason why our methods would not work on non-MSM stories, if they are discussed on the online social networks. However, this is harder to evaluate given the lack of a convenient ground-truth.

We filtered Twitter using relevant keywords and hashtags to collect tweets around three events: the 2012 "Super Tuesday" primaries, part of the presidential nomination race of the US Republican Party; the 2012 FA Cup final, the climax to the English football season; and the 2012 US presidential election, an event of global significance.

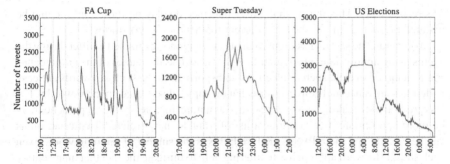

Fig. 1 Twitter activity during events (tweets per minute). For the FA Cup, the peaks correspond to start and end of the match and the goals. For the two political collections, the peaks correspond to the main result announcements

In each case, we reviewed the published MSM accounts of the events and chose a set of stories that were significant, time-specific, and represented on Twitter. For example, we ignored general reviews of the state of US politics (not time-specific), and quotes from members of the public (not significant events).

Using MSM sources presents its own problems however. Each MSM source has its own policy for selecting and sharing stories, most obviously being national biases (e.g. UK outlets tend to emphasise UK stories). To get detailed accounts of the events of interest, we relied on "live blogs" produced by various MSM outlets [41]; again, many events are not covered by live blogs, potentially introducing a further bias into our selection of topics. Of course the choice of *which* MSM sources to use is critical and to some extent subjective. We chose MSM outlets that have an excellent reputation for timely and reliable reporting, primarily the BBC and the Wall Street Journal.

For each target topic, we identified around 5–7 keywords that defined the story and used these to measure recall and precision, as discussed below. Some examples are shown in the first two columns of Table 4. We also defined several "forbidden" keywords. A topic was only considered as successfully recalled if all of the "mandatory" terms were retrieved and *none* of the "forbidden" terms. The aim was to avoid producing topics such as "victory Romney Paul Santorum Gingrich Alaska Georgia" that convey no information about who won or where; or "Gingrich wins", which is too limited to define the story because it doesn't name the state where the victory occurred. Similarly, when detecting events during a football match, topics labels such as "Liverpool Chelsea goal" or just "goal" are not useful.

Figure 1 shows the frequency of tweets collected over time, with further details in Ref. [2]. We have made all the data freely available, including the ground truth topics.[1] The three data sets differ in the rates of tweets, determined by the popularity of the topic and the choice of filter keywords. The mean tweets per minute (tpm) were: Super Tuesday, 832 tpm; FA Cup, 1293 tpm; and US elections, 2209 tpm. For a slot size of 1500 tweets these correspond to a "topic refresh rate" of 108, 70 and

---

[1] http://www.socialsensor.eu/results/datasets/72-twitter-tdt-dataset.

41 s respectively. This means that a user interface displaying these topics could be updated every 1–2 min to show the current top-10 (or top-m) stories.

To generate the sets of Tweets used in the evaluation, we crawled Twitter during the events using appropriate sets of filter keywords, such as the names of the participants. For timetabled events, such as elections and sports fixtures, such keywords are easy to define in advance and help to ensure that the topics discovered are newsworthy. For less predictable breaking news stories, such as natural disasters, other approaches may be more appropriate. For example, a list of reliable, news-related Twitter accounts can be created and their Tweets analysed; this is the subject of ongoing work. Even such straightforward approaches to filtering help to mitigate the fact that many bursty topics on Twitter would not usually be considered newsworthy [8].

We ran the topic detection algorithm on each data set. This produced a ranked list of topics, each defined by a set of terms (i.e. n-grams). For our evaluation, we focus on the recall of the top m topics ($1 \leq m \leq 10$) at the time each ground-truth story emerges. For example, if a particular story was being discussed in the mainstream media from 10:00–10:15, then we consider the topic to be recalled if the system ranked it in the top m at any time during that period.

The automatically detected topics were compared to the ground truth (comprising 22 topics for Super Tuesday; 13 topics for FA Cup final; and 64 topics for US elections) using three metrics:

- Topic recall: Percentage of ground truth topics that were successfully detected. A topic was considered successfully detected if the automatically produced set of words contained all mandatory keywords for it (and none of the forbidden terms, if defined).
- Keyword precision: Percentage of correctly detected keywords out of the total number of keywords for all topics detected by the algorithm in the slot.
- Keyword recall: Percentage of correctly detected keywords divided by the total number of ground truth keywords (excluding forbidden keywords) in the slot. One key difference between "topic recall" and "keyword recall" is that the former is a user-centred evaluation metric, as it considers the power of the system at retrieving and displaying to the user stories that are meaningful and coherent, as opposed to retrieving only some keywords that are potentially meaningless in isolation.

Note that we do not attempt to measure topic precision as this would need an estimate of the total number of newsworthy topics at any given time, in order to verify which (and how many) of the topics returned by our system were in fact newsworthy. This would require an exhaustive manual analysis of MSM sources to identify every possible topic (or some arbitrary subset), which is infeasible. One option is to compare detected events to some other source, such as Wikipedia, to verify the significance of the event [29], but Wikipedia does not necessarily correspond to particular journalists' requirements regarding newsworthiness and does not claim to be complete. The scores reported below were automatically computed by an evaluation script. However, to ensure the reliability of results, we conducted several rounds of manual evaluation of results and confirmed their agreement with the automatically produced ones.

Table 1 Topic recall for different slot sizes (with hierarchical clustering)

Slot size (tweets)	500	1000	1500	2000	2500
Super tuesday	**0.773**	0.727	0.682	0.545	0.682
FA cup	0.846	0.846	**0.923**	**0.923**	**0.923**
US elections	0.750	0.781	**0.844**	0.734	0.766
Weighted mean	0.77	0.78	**0.82**	0.72	0.77

4.2 Results

Table 1 shows the effect on topic recall when varying the slot size, with the same total number of topics in the evaluation for each slot size. The mean is weighted by the number of topics in the evaluation for each slot size. The mean is weighted by the number of topics in the ground truth for each set, giving greater importance to larger test sets. Overall, using very few tweets produces slightly worse results than with larger slot sizes (e.g. 1500 tweets), presumably as there is too little information in such a small collection. Slightly better results for the Super Tuesday set occur with fewer tweets; this could be due to the slower tweet rate in this set. Note that previous experiments [23] showed that including 3-g improves recall compared to just using 1- and 2-g, but adding 4-g provides no extra benefit, so here we use 1-, 2- and 3-g phrases throughout.

Lastly, we compared the results of combining different clustering techniques with different topic ranking techniques (see Fig. 2). We conclude that the hierarchical clustering performs well despite the weakness discussed above (i.e. each n-gram is assigned to only one cluster), especially in FA Cup dataset. Also, the use of weighted topic-length ranking technique improves topic recall with hierarchical clustering in the political data sets.

The Apriori algorithm performs quite well in combination with the weighted topic length ranking technique (note that this ranking technique was specially created for the "partial" membership clustering techniques). We see that the Apriori algorithm in combination with the maximum n-gram $df - idf_t$ ranking technique produces slightly worse results, as this ranking technique does not produce diverse topics for the first results (from top 1 to top 10, in our case) as we mentioned earlier.

Turning to the EM Gaussian mixture model results, we see that this method works very well on the FA Cup final and US elections data sets. Despite being a "partial" membership clustering technique, the use of weighted topic length ranking technique does not make any representative difference, even its performance is worse in Super Tuesday dataset. Further work is needed to test this.

Table 2 summarises the results of the three clustering methods and the two ranking methods across all three data sets. The weighted-mean scores show that for the three clustering methods, ranking by the length of the topic is more effective than ranking by each topic's highest $df - idf_t$ score. We can see that for the FA Cup set, the Hierarchical and GMM clustering methods are the best ones in combination with

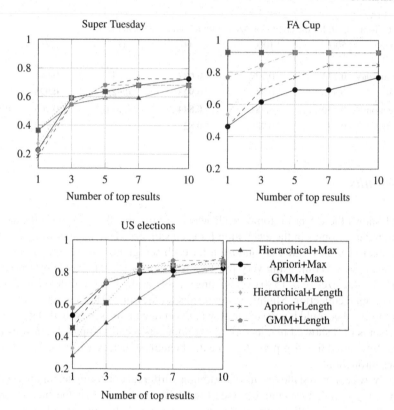

Fig. 2 Topic recall for different clustering techniques in the Super Tuesday, FA Cup and US elections sets (slot size = 1500 tweets)

Table 2 Normalised area under the curve for the three datasets combining the different clustering and topic ranking techniques (1500 tweets per slot)

Ranking	Max. n-gram $df - idf_t$			Weighted topic-length		
Clustering	Hierar.	Apriori	GMM	Hierar.	Apriori	GMM
FA cup	**0.923**	0.677	**0.923**	0.861	0.754	0.892
Super tuesday	0.573	0.605	0.6	0.591	**0.614**	0.586
US elections	0.627	0.761	0.744	0.761	0.772	**0.797**
Weighted mean	0.654	0.715	0.735	0.736	0.734	**0.763**

the maximum n-gram $df - idf_t$ ranking technique. For Super Tuesday and US Elections data sets, "partial" membership clustering techniques (Apriori and GMM, respectively) perform the best in combination with weighted topic length ranking technique, as expected.

Finally, Table 3 shows more detailed results, including keyword precision and recall, for the best combinations of clustering and topic ranking methods of the three datasets when the top five results are considered per slot. In addition, Table 4 shows some examples of ground truth and BNgram detected topics and tweets within the corresponding detected topics for all datasets.

Table 3 Best results for the different datasets after evaluating top 5 topics per slot

Method	T-REC@5	K-PREC@5	K-REC@5
Super tuesday			
Apriori+Length	0.682	0.431	0.68
GMM+Length	0.682	0.327	0.753
FA cup			
Hierar.+Max	0.923	0.337	0.582
Hierar.+Length	0.923	0.317	0.582
GMM+Max	0.923	0.267	0.582
GMM+Length	0.923	0.162	0.673
US elections			
GMM+Max	0.844	0.232	0.571

T-REC, K-PREC, and K-REC refers to topic-recall and keyword-precision/recall respectively

Table 4 Examples of the mainstream media topics, the target keywords, the topics extracted by the $df - idf_t$ algorithm, and example tweets selected by our system from the collections

Target topic	Ground truth keywords	Extracted keywords	Example tweet
Newt Gingrich says "Thank you Georgia! It is gratifying to win my home state so decisively to launch our March Momentum"	Newt Gingrich, Thank you, Georgia, March, Momentum, gratifying	launch, March, Momentum, decisively, thank, Georgia, gratifying, win, home, state, #MarchMo, #250gas, @newtgingrich	@Bailey_Shel: RT @newtgingrich: Thank you Georgia! It is gratifying to win my home state so decisively to launch our March Momentum. #MarchMo #250gas
Salomon Kalou has an effort at goal from outside the area which goes wide right of the goal	Salomon Kalou, run, box, mazy	Liverpool, defence, before, gets, ambushed, Kalou, box, mazy, run, @chelseafc, great, #cfcwembley, #facup, shoot	@SharkbaitHooHa_: RT @chelseafc: Great mazy run by Kalou into the box but he gets ambushed by the Liverpool defence before he can shoot #CFCWembley #FACup
US President Barack Obama has pledged "the best is yet to come", following a decisive re-election victory over Republican challenger Mitt Romney	Obama, best, come	America, best, come, United, States, hearts, #Obama, speech, know, victory	@northoaklandnow: "We know in our hearts that for the United States of America, the best is yet to come," says #Obama in victory speech

5 Applications

In this section, we discuss several specific applications of our clustered BNgram approach. These go some way to demonstrate the robustness of the algorithm and explore how it can be applied to sports, subjective event summarization and rolling 24-h news.

5.1 Finding Events in Football Matches

Earlier in this chapter (Sect. 4), we described analysis of the 2012 FA Cup Final. In a recent paper [9], we also compared this with the 2013 Final, a match between Manchester City and Wigan Athletic. We used this as a further means to evaluate the automated topic detection system, once again using a ground truth derived from mainstream media.

In this case, as well as detecting topics, we also attempted to identify the team that each Twitter user supported, or to recognise their neutrality. For each user, we counted the number of times they mentioned each team in all their tweets. An initial manual inspection showed that fans tend to use their team's standard abbreviation as a hashtag (e.g. #CFC or #MCFC) a great deal more often than any other teams', irrespective of sentiment. We therefore define a fan's degree of support for one team as how many more times that team's abbreviation is mentioned by the user compared to their second-most mentioned team. Here, we include as "fan" any user with a degree of two or more and treat everyone else as neutral. A manual evaluation indicated that this approach identifies which team is supported for over 90 % of the tweeters but occasionally mis-identifies neutral reporters as supporting one team. We believe this could be improved if we extended the analysis over several matches to build up more evidence for support.

We chose a total of 25 events from the two matches that were reported by the mainstream media (specifically the BBC commentaries). Of these our system found around 75–90 % of the events. The variation is likely due to the different nature of the two matches considered and the volume of tweets generated. Repeating over more matches would give us a clearer indication of quality, but clearly the algorithm does find a large number of the most important events.

Other systems have been proposed to discover events within sporting fixtures, but these typically are designed to find only events of pre-defined classes, such as goals or bookings [42, 43]. In contrast, ours is agnostic about the specific nature of the event, relying only on shifts in the word-use used by multiple users to describe it.

We also showed that fans of each team tended to give biased, subjective views of the events, as would be expected. We explored this further in our next paper (see Sect. 5.2).

5.2 Subjective Summarization of Sporting Events

In most journalism, there is an aim of objectively summarizing the events and pre-
senting them from a neutral point of view, although there is some debate about how
much this is really possible, or even desirable [11]. However, when producing such a
neutral point of view, there is a risk that the distinct opinions expressed by different
groups get lost in the mix. This range of opinions is often of interest to journalists
and to news consumers, as it reflects diversity. In a democracy, it is important that
different arguments are presented and considered, and this may effect people's opinions. Related work on automated document summarization has sometimes attempted
to distinguish and summarize diverse opionions [20], but this is rare. In sports, fans
may rarely change their allegiance to one team or another, but it is still interesting to consider the range of opinions expressed. Our work here can be seen as a
step towards subjective event summarization, summarizing messages from specific,
distinct points-of-view.

Having shown that we could (a) identify key events of football matches, and (b)
identify which team each tweeter supported, we then combined these two methods
into a subjective event summarization tool, as described in a recent paper [10]. We
used the same method as before to estimate which team each tweeter supports, if any
(Sect. 5.1), and the same BNgram topic detection methods. This time, we also tracked
the relatively objective, neutral mainstream media comments from the BBC's live
text-based commentaries. For each slot, we used our BNgram algorithm to select
up to 10 topics. We then compared these to the corresponding BBC commentary,
using a simple cosine similarity, and selected the most similar. In this way, we could
discover what each set of fans were *subjectively* saying about the events that were
objectively most important. As an alternative to the BBC commentary, we could have
used the BNgram algorithm on the entire collection of tweets, thus incorporating the
view of both sets of fans and the many neutral observers, when determining the
"objective" event list.

A distinct but related approach is to identify reliable "reporters" of events, such
as people watching a football match who also provide regular, accurate tweets about
it [21]. In common with several event-detection approaches [42, 43], they rely on
spikes in the overall activity of message streams to identify events, unlike our work
which only needs the frequency of terms to shift within a (potentially) unchanging
volume of messages.

Although not strictly related to topic detection, we have also analysed tweets sent
by fans of different teams during English Premier League matches. In that work [7],
we focussed on the use of swearing in tweets and how curse words are used to express
sentiment, both positive and negative. This contrasts with an assumption common
to much sentiment analysis research, that swearing is more typically negative or
sarcastic, and rarely positive [22, 25].

5.3 Real-Time Topic Detection for 24 Hours of News

Our previous studies described above have all focussed on specific, pre-specified events: the Super Tuesday primaries; the 2012 US Presidential Elections; and the 2012 and 2013 FA Cup Finals. While very useful as benchmarks, there is a risk that methods developed to analyse such specific events may fail to generalise to the wider case of finding newsworthy stories during a typical 24-h news cycle. To test our approach in this scenario, we entered the 2014 Social News On the Web (SNOW) Data Challenge[2] [24]. This challenge is held in conjunction with the 23rd International World Wide Web Conference (WWW 2014).

The task of this challenge is to retrieve newsworthy stories or topics for multiple timeslots over 24 h, where each timeslot is 15 min long. The required format of each topic includes a human-readable label, a set of the most representative keywords, a set of tweets that are related to the story and links to any relevant images from the tweets. The tweets identified for each event could be from the corresponding timeslot or any earlier one (but not later), to simulate a real-time scenario.

As the guidelines of the challenge show, the extracted topics were evaluated on several dimensions, namely: precision and recall, readability, coherence, relevance and diversity. Further details can be found in the official description of this challenge [31].

Regarding our BNgram approach, we modified the strategy slightly to select bursty terms after analysis of topics produced during previous experiments. Bigrams, trigrams, entities, hashtags and URLs were considered as terms in the SNOW experiments. The Apriori algorithm (see Sect. 3.2) was used for the clustering algorithm. In addition, we considered temporal windows (i.e. timeslots) instead of using a fixed number of tweets per slot, and used two previous timeslots for the penalization of common terms. The final formula to compute $df - idf$ scores was (mostly based on Eq. 1):

$$df - idf_{ti} = (df_{ti} + 1) \cdot \frac{1}{\log \left(\frac{\sum_{j=i}^{s} df_{t(i-j)}}{s} + 1 \right) + 1}. \tag{3}$$

where $s = 2$ in the experiments as before.

To populate topics with tweets, our approach creates a query based on the most representative terms to retrieve the associated tweets to the story. In addition, replies to the previous tweets are also considered as they can add further details of the story. The main reasons to include them is that they are not text-query dependant and add a wider range of people's view in many cases. However, we believe a filtering process should be considered for these replies, as we detected many spam replies, such as advertising links.

Our topic label approach here is based on the selection of the most representative tweet from the set of topic tweets, following some recent advice that headlines

---

[2]http://www.snow-workshop.org/.

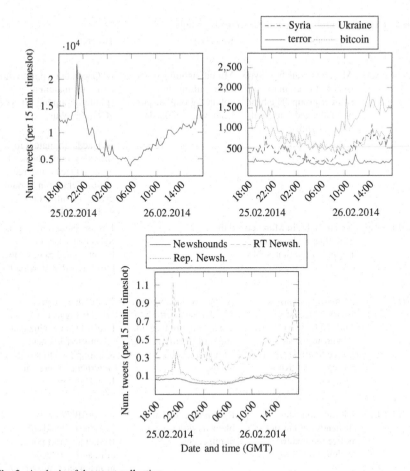

Fig. 3 Analysis of data test collection

increasingly resemble tweets.[3] Therefore, the tweet containing the greatest number of topic terms and duplicates (e.g. retweets) is selected. Its text, is then "cleaned" (by removing redundant user mentions, URLs and abbreviations such as RT and MT) to make it more readable, and is then used as the topic title or label.

The final topics are ranked by their bursty scores where each score is the maximum $df - idf_t$ value of their constituent terms (see Sect. 3.3). Our assumption is that the the most popular term from each topic represents the core of the topic and diverse topics are detected by the algorithm as the collection is not event-based.

Our final test data collection was composed of 901,895 tweets and stored in Solr after filtering out the non-English tweets. We extracted entities from each tweet using the Stanford NLP library [14], and created links between replies and retweets with their original tweets.

---

[3]http://perryhewitt.com/5-lessons-buzzfeed-harvard/.

Table 5 Examples of topics about the tracking keywords

Timeslot	Topic label	Keywords	Tweets
Syria			
25/2/14 20:30	Al Qaeda branch in Syria issues ultimatum to splinter group: The head of an al Qaeda-inspired militia fighting	Militia, fighting, branch, issues, ultimatum, splinter, group, inspired, head, syria, al qaeda	Al Qaeda branch in Syria issues ultimatum to splinter group http://t.co/gQDm0p7Wur
			Al Qaeda ultimatum to splinter group: The head of an al Qaeda-inspired militia fighting in Syria is giving a... http://t.co/9KFu1CG1F6
26/2/14 00:15	Jordan Bahrain Morocco Syria Qatar Oman Iraq Egypt United States 346	346	Jordan Bahrain Morocco Syria Qatar Oman Iraq Egypt United States 346 http://t.co/RjZAwwMJ95
Terror			
26/2/14 4:00	25 marines to arrest 'worlds biggest drug lord' El Chapo Guzman 73 anti-terror-squad police to arrest 'Internet entrepreneur' Kim Dotcom	Arrest, worlds, biggest, 25, marines, terror, squad, amp, police, 73, anti, drug, lord, internet, entrepren,el chapo guzman	25 marines to arrest 'worlds biggest drug lord' El Chapo Guzman 73 anti-terror-squad & police to arrest 'Internet entrepreneur' Kim Dotcom
Ukraine			
26/2/14 10:15	Ukraine minister disbands Berkut riot police blamed for violence—CNN	Riot, police, disbands, blamed, violence, ukraine, cnn	RT @BBCWorld: Ukraine disbands elite Berkut anti-riot police unit, acting interior minister says http://t.co/5GqM6jjryu
			RT @cnnbrk: Ukraine has disbanded a riot police force used against anti-government protesters, acting interior minister said
			RT @BBCGavinHewitt: In Ukraine the Berkut special police units blamed for most of the shootings have been disbanded

(continued)

Table 5 (continued)

Timeslot	Topic label	Keywords	Tweets
Bitcoin			
25/2/14 20:00	Mt. Goxs Demise Marks The End of Bitcoins First Wave Of Entrepreneurs	Demise, marks, end, first, wave, gox, bitcoin, entrepreneurs	Mt. Gox's Demise Marks The End of Bitcoin's First Wave Of Entrepreneurs http://t.co/gIKKP3RLQn by @kimmaicutler
			Mt. Gox's Demise Marks The End of Bitcoin's First Wave Of... http://t. co/X7iUKN3Vsv #eCommerce #Finance #Startups #TC #techcrunch #tech
			#SuryaRay #Surya #SuryaRay #Surya Mt. Gox's Demise Marks The End of Bitcoin's... http:// t.co/csX2dB26w4 @suryaray @suryaray @suryaray3

Figure 3a shows the distribution of tweets per 15-min timeslot for this final collection going from 18:00 25/2/2014 to 18:00 26/2/2014. Note that the high peak shown from 20:00 to 22:00 on 25/2/2014 corresponds to tweets (mainly retweets and replies, as shown in Fig. 3c) related to several Champions League football matches that were taking place at that time. This is because there were some sport commentators in the list of accounts used to select the tweets (as provided by the SNOW challenge organizers). There are no clear peaks during that period related to the keywords provided for tracking, as shown by Fig. 3b, as these are not football related. Finally, the activity goes down overnight as most of the Twitter accounts being followed are UK-based, so it is more likely they were inactive during these hours.

Table 5 shows some representative topics associated to the tracked keywords, giving some indication of the quality of the stories found.

The official evaluation results of our method in the SNOW Data Challenge are included in Papadopoulos et al. [31]. Overall, our submission was placed second out of the eleven teams from round the world that completed the challenge. The winning team of Ifrim et al. also used our BNgram approach to rank and filter topics, alongside more aggressive pre-processing and filtering methods [19]. While neither team found every one of the target topics defined by the challenge organizers, the fact that the two best-placed teams used variations of the same BNgram algorithm strongly suggests that this is a robust and flexible tool for detecting topics in Twitter streams.

6 Conclusions

In Sect. 4, we presented our main findings regarding the power of the BNgram algorithm. If we compare the results between the three main collections, one difference is particularly striking: the topic recall is far higher for football (over 90 %) than for politics (around 60–80 %; Table 2). This is likely to reflect the different nature of conversations about the events. Topics within a live sports event tend to be transient: fans care (or at least tweet) little about what happened 5 min ago; what matters is what is happening "now". This is especially true during key events, such as goals, as also discussed in Sects. 5.1 and 5.2. In politics, conversations and comments tend to spread over hours (or even days) rather than minutes. This means that sports-related topics tend to occur over a much narrower window, with less overlapping chatter. In politics, several different topics are likely to be discussed at the same time, making this type of trend detection much harder. Looking back at the distribution of the tweets over time (Fig. 1), we can see clear spikes in the FA Cup graph, each corresponding to a major event (kick-off, goals, half-time, full-time etc.). No such clarity is in the politics graphs, which instead is best viewed as many overlapping trends.

This difference is reflected in the way that major news stories often emerge: an initial single, focussed story emerges but is later replaced with several potentially overlapping sub-stories covering different aspects of the story. Our results suggest that a dynamic approach may be required for newsworthy topic detection, finding an initial clear burst and subsequently seeking more subtle and overlapping topics. The specific applications we described included analysis of 24 h of news-related tweets (Sect. 5.3). In this work, we saw more clearly that news stories tend to emerge over time, to overlap greatly and to have multiple angles. As more details emerge around breaking news stories, it becomes increasingly important to go further than topic detection and to start identifying links between topics.

Recently, Twitter has been actively increasing its ties to television.[4] Broadcast television and sporting events share several common features: they occur a pre-specified times; they attract large audiences; and they are fast-paced. These features all allow and encourage audience participation in the form of sharing comments and holding discussions during the events themselves, such that the focus of the discussion is constantly moving with the event itself. Potentially, this can allow targeted time-sensitive promotions and advertising based on topics currently receiving the most attention. Facebook and other social media are also competing for access to this potentially valuable "second screen" [16]. Television shows are increasingly promoting hashtags in advance, which may make collecting relevant tweets more straightforward. One potential approach to help with this is a "visual backchannel" [13] that allows users to visualize and make sense of masses of streaming information, and this does could be enhanced with improved topic detection and clustering.

---

[4]"Twitter & TV: Use the power of television to grow your impact" https://business.twitter.com/twitter-tv.

Even if topic detection for news requires slightly different methods or parameters when compared to detecting sporting and live television events, all these areas have substantial and growing demand.

Acknowledgments This work is supported by the SocialSensor FP7 project, partially funded by the EC under contract number 287975. We wish to thank Nic Newman and Steve Schifferes of the Department of Journalism, City University London and Andrew MacFarlane of the Department of Computer Science, City University London, for their invaluable advice.

References

1. Agrawal, R., Srikant, R., et al.: Fast algorithms for mining association rules. In: Proceedings of the 20th International Conference on Very Large Data Bases, VLDB, vol. 1215, pp. 487–499 (1994)
2. Aiello, L., Petkos, G., Martin, C., Corney, D., Papadopoulos, S., Skraba, R., Göker, A., Kompatsiaris, I., Jaimes, A.: Sensing trending topics in Twitter. IEEE Trans. Multimedia **15**(6), 1268–1282 (2013). doi:10.1109/TMM.2013.2265080
3. Alvanaki, F., Sebastian, M., Ramamritham, K., Weikum, G.: Enblogue: emergent topic detection in Web 2.0 streams. In: Proceedings of the 2011 International Conference on Management of Data, pp. 1271–1274. ACM (2011)
4. Becker, H., Naaman, M., Gravano, L.: Beyond trending topics: real-world event identification on Twitter. In: Proceedings of the Fifth International AAAI Conference on Weblogs and Social Media (ICWSM11) (2011)
5. Benhardus, J.: Streaming trend detection in Twitter. National Science Foundation REU for Artificial Intelligence, Natural Language Processing and Information Retrieval, University of Colorado (2010)
6. Blei, D.M., Ng, A.Y., Jordan, M.I.: Latent Dirichlet Allocation. J. Mach. Learn. Res. **3**, 993–1022 (2003)
7. Byrne, E., Corney, D.: Sweet FA: sentiment, swearing and soccer. In: ICMR2014 1st Workshop on Social Multimedia and Storytelling. Glasgow, UK (2014)
8. Castillo, C., Mendoza, M., Poblete, B.: Information credibility on Twitter. In: Proceedings of the 20th International Conference on World Wide Web, pp. 675–684. ACM (2011)
9. Corney, D., Martin, C., Göker, A.: Spot the ball: detecting sports events on Twitter. In: ECIR 2014, pp. 449–454. Amsterdam, Holland (2014)
10. Corney, D., Martin, C., Göker, A.: Two sides to every story: Subjective event summarization of sports events using Twitter. In: ICMR2014 1st Workshop on Social Multimedia and Storytelling. Glasgow, UK (2014)
11. Cunningham, B.: Re-thinking objectivity. Columbia. Journalism Rev. **42**(2), 24–32 (2003)
12. Dempster, A.P., Laird, N.M., Rubin, D.B.: Maximum likelihood from incomplete data via the EM algorithm. J. R. Stat. Soc. Ser. B (Methodological), 1–38 (1977)
13. Dork, M., Gruen, D., Williamson, C., Carpendale, S.: A visual backchannel for large-scale events. IEEE Trans. Vis. Comput. Graph. **16**(6), 1129–1138 (2010)
14. Finkel, J.R., Grenager, T., Manning, C.: Incorporating non-local information into information extraction systems by Gibbs sampling. In: Proceedings of the 43rd Annual Meeting on Association for Computational Linguistics, ACL '05, pp. 363–370. Stroudsburg, PA, USA (2005). doi:10.3115/1219840.1219885
15. Fraley, C., Raftery, A.E.: How many clusters? Which clustering method? Answers via model-based cluster analysis. Comput. J. **41**(8), 578–588 (1998)
16. Goel, V., Stelter, B.: Social networks in a battle for the second screen. The New York Times (2013). http://www.nytimes.com/2013/10/03/technology/social-networks-in-a-battle-for-the-second-screen.html. Accessed 24 Mar 2014

17. Hall, M., Frank, E., Holmes, G., Pfahringer, B., Reutemann, P., Witten, I.H.: The WEKA data mining software: an update. ACM SIGKDD Explor. Newsl. **11**(1), 10–18 (2009)
18. He, D., Göker, A., Harper, D.: Combining evidence for automatic web session identification. Inf. Process. Manage. **38**(5), 727–742 (2002)
19. Ifrim, G., Shi, B., Brigadir, I.: Event detection in Twitter using aggressive filtering and hierarchical tweet clustering. In: Proceedings of the SNOW 2014 Data Challenge (2014)
20. Ku, L.W., Lee, L.Y., Wu, T.H., Chen, H.H.: Major topic detection and its application to opinion summarization. In: 28th ACM SIGIR Conference, pp. 627–628. ACM (2005)
21. Kubo, M., Sasano, R., Takamura, H., Okumura, M.: Generating live sports updates from Twitter by finding good reporters. In: IEEE/WIC/ACM International Joint WI-IAT Conferences, vol. 1, pp. 527–534. IEEE (2013)
22. Liu, B.: Sentiment analysis and subjectivity. In: N. Indurkhya, F.J. Damerau (eds.) Handbook of Natural Language Processing, 2nd edn. Chapman & Hall, Boca Raton (2010)
23. Martin, C., Corney, D., Göker, A.: Finding newsworthy topics on Twitter. IEEE Comput. Soc. Spec. Tech. Community Soc. Netw. E-Letter **1**(3) (2013)
24. Martin, C., Göker, A.: Real-time topic detection with bursty n-grams: RGU's submission to the 2014 SNOW challenge. In: Proceedings of the SNOW 2014 Data Challenge (2014)
25. Maynard, D., Bontcheva, K., Rout, D.: Challenges in developing opinion mining tools for social media. In: Proceedings of @NLP can u tag #usergeneratedcontent?! Workshop at LREC 2012. Turkey (2012)
26. Murtagh, F.: A survey of recent advances in hierarchical clustering algorithms. Comput. J. **26**(4), 354–359 (1983)
27. Newman, N.: #ukelection2010, mainstream media and the role of the internet. Reuters Institute for the Study of Journalism working paper (2010)
28. Newman, N.: Mainstream media and the distribution of news in the age of social discovery. Reuters Institute for the Study of Journalism working paper (2011)
29. Osborne, M., Petrovic, S., McCreadie, R., Macdonald, C., Ounis, I.: Bieber no more: First story detection using Twitter and Wikipedia. In: SIGIR 2012 Workshop on Time-aware Information Access (2012)
30. Ozdikis, O., Senkul, P., Oguztuzun, H.: Semantic expansion of hashtags for enhanced event detection in Twitter. In: Proceedings of VLDB 2012 Workshop on Online Social Systems (2012)
31. Papadopoulos, S., Corney, D., Aiello, L.M.: SNOW 2014 data challenge: Assessing the performance of news topic detection methods in social media. In: Proceedings of the SNOW 2014 Data Challenge (2014)
32. Petrovic, S., Osborne, M., Lavrenko, V.: Streaming first story detection with application to Twitter. In: Proceedings of Human Language Technologies: 2010 Conference of NAACL, vol. 10 (2010)
33. Petrovic, S., Osborne, M., Lavrenko, V.: Using paraphrases for improving first story detection in news and Twitter. In: Proceedings of HTL12 Human Language Technologies, pp. 338–346 (2012)
34. Phuvipadawat, S., Murata, T.: Breaking news detection and tracking in Twitter. In: Proceedings of the 2010 IEEE/WIC/ACM International Conference on Web Intelligence and Intelligent Agent Technology, vol. 3, pp. 120–123 (2010)
35. Phuvipadawat, S., Murata, T.: Detecting a multi-level content similarity from microblogs based on community structures and named entities. J. Emerg. Technol. Web Intell. **3**(1), 11–19 (2011)
36. Ratkiewicz, J., Conover, M., Meiss, M., Gonçalves, B., Flammini, A., Menczer, F.: Detecting and tracking political abuse in social media. In: Proceedings of the International Conference on Weblogs and Social Media (ICWSM) (2011)
37. Sayyadi, H., Hurst, M., Maykov, A.: Event detection and tracking in social streams. In: Proceedings of International Conference on Weblogs and Social Media (ICWSM) (2009)
38. Schifferes, S., Newman, N., Thurman, N., Corney, D., Göker, A., Martin, C.: Identifying and verifying news through social media. Digital Journalism (2014). doi:10.1080/21670811.2014.892747

39. Shamma, D., Kennedy, L., Churchill, E.: Peaks and persistence: modeling the shape of microblog conversations. In: Proceedings of the ACM 2011 conference on Computer Supported Co-operative Work, pp. 355–358. ACM (2011)
40. Spärck, J.K.: A statistical interpretation of term specificity and its application in retrieval. J. Documentation **28**(1), 11–21 (1972)
41. Thurman, N., Walters, A.: Live blogging—digital journalism's pivotal platform? A case study of the production, consumption, and form of live blogs at Guardian.co.uk. Digital Journalism **1**(1), 82–101 (2013)
42. van Oorschot, G., van Erp, M., Dijkshoorn, C.: Automatic extraction of soccer game events from Twitter. In: Proceedings of the Workshop on Detection, Representation, and Exploitation of Events in the Semantic Web (2012)
43. Zhao, S., Zhong, L., Wickramasuriya, J., Vasudevan, V.: Human as real-time sensors of social and physical events: A case study of Twitter and sports games. arXiv preprint arXiv:1106.4300 (2011)

Sentiment Analysis Using Domain-Adaptation and Sentence-Based Analysis

Gizem Gezici, Berrin Yanikoglu, Dilek Tapucu and Yücel Saygın

Abstract Sentiment analysis aims to automatically estimate the sentiment in a given text as positive, objective or negative, possibly together with the strength of the sentiment. Polarity lexicons that indicate how positive or negative each term is, are often used as the basis of many sentiment analysis approaches. Domain-specific polarity lexicons are expensive and time-consuming to build; hence, researchers often use a general purpose or domain-independent lexicon as the basis of their analysis. In this work, we address two sub-tasks in sentiment analysis. We apply a simple method to adapt a general purpose polarity lexicon to a specific domain [1]. Subsequently, we propose and evaluate new features to be used in a word polarity based approach to sentiment classification. In particular, we analyze sentences as the first step for estimating the overall review polarity. We consider different aspects of sentences, such as length, purity, irrealis content, subjectivity, and position within the opinionated text. This analysis is then used to find sentences that may convey better information about the overall review polarity. We use a subset of hotel reviews from the TripAdvisor database [2] to evaluate the effect of sentence-level features on sentiment classification. Then, we measure the performance of our sentiment analysis engine using the domain-adapted lexicon on a large subset of the TripAdvisor database.

G. Gezici · B. Yanikoglu (✉) · Y. Saygın
Faculty of Engineering and Natural Sciences, Sabanci University, 34956 Istanbul, Turkey
e-mail: berrin@sabanciuniv.edu

G. Gezici
e-mail: gizemgezici@sabanciuniv.edu

Y. Saygın
e-mail: ysaygin@sabanciuniv.edu

D. Tapucu
Department of Computer Engineering, Izmir Institute of Technology, 35430 Izmir, Turkey
e-mail: dilektapucu@sabanciuniv.edu

© Springer International Publishing Switzerland 2015
M.M. Gaber et al. (eds.), *Advances in Social Media Analysis*,
Studies in Computational Intelligence 602,
DOI 10.1007/978-3-319-18458-6_3

45

1 Introduction

Sentiment analysis aims to extract the subjectivity and strength of the opinions indicated in a given text; which together indicate its *semantic orientation*. For instance, a given word or sentence in a specific context, or a review about a particular product can be analyzed to determine whether it is objective or subjective, together with the *polarity* of the opinion. The polarity itself can be indicated categorically as positive, objective or negative; or numerically, indicating the *strength of the opinion* in a canonical scale.

Automatic extraction of the sentiment can be very useful in analyzing what people think about specific issues or items, by analyzing large collections of textual data sources such as personal blogs, review sites, and social media. Commercial interest to this problem has shown to be strong, with companies showing interest to public opinion about their products; and financial companies offering advice on general economic trend by following the sentiment in social media [3]. In the remainder of this chapter, we use the terms "document", "review" and "text" interchangeably, to refer to the text whose sentiment polarity or opinion strength is to be estimated.

Two main approaches for sentiment analysis are defined in the literature: one approach is called *lexicon-based* [4] and the other is based on *supervised learning* [5]. The lexicon-based approach calculates the semantic orientation of a given text from the polarities of the constituent words or phrases [4], obtained from a lexicon such as the SentiWordNet [6]. In this approach, different features of the text may be extracted from word polarities [7], such as average word polarity, or the number of subjective words, but the distinguishing aspect is that there is no supervised learning. Furthermore, the text is often treated as a *bag-of-words*; in other words, features are obtained from constituent words without keeping track of the location of those words. Alternatives to the bag-of-word approach are also possible, where word polarities of the first sentence etc. are calculated separately [8]. Furthermore, as words may have different connotations in different domains (e.g. the word "small" has a positive connotation in cell phone domain; while it is negative in hotel domain), one can use a domain-specific lexicon whenever available. The widely used SentiWordNet [6] and SenticNet [9] are two widely known domain-independent lexicons.

Supervised learning approaches use machine learning techniques to establish a model from an available corpus of reviews with associated labels. For instance in [5, 10], researchers use the Naive Bayes algorithm to separate positive reviews from negative ones by learning the conditional probability distributions of the considered features in the two classes. Note that in supervised learning approaches, a polarity lexicon may still be used to extract features of the text, such as average word polarity and the number of positive words etc., that are later used in a learning algorithm. Alternatively, in some supervised approaches the lexicon is not needed. For instance in the Latent Dirichlet Analysis (LDA) approach, a training corpus is used to learn the probability distributions of topic and word occurrences in the different categories (e.g. positive or negative sets of reviews) and a new text is classified according to its likelihood of coming from these different distributions [11, 12]. While supervised approaches are typically more successful than lexicon-based ones, collecting a large amount of labelled, domain-specific data can be a problem.

In this work, we present a supervised learning approach to sentiment analysis, addressing two sub-tasks of the problem. First, we apply a simple domain-adaptation method proposed in [1] to adapt a domain-independent polarity lexicon to a specific domain. We show that even changes in the polarity of a small number of words affect the overall accuracy by a few percent. Next, we propose a sentence-based analysis of the review sentiment, using the updated polarity lexicon for feature extraction. While word-level polarities provide a simple, yet effective method for estimating a review's polarity, the gap from word-level polarities to review polarity is too big. The use of sentence-based analysis is aimed to bridge this gap.

The remainder of this chapter is organized as follows. Section 2 provides an overview of related work. Section 3 proposes the adaptation process of a domain-independent lexicon. Section 4 describes our sentence-based sentiment analysis approach that forms the main contribution in this work. Section 5 presents the learning module and Sect. 6 reports experimental results. Finally, in Sect. 7 we conclude and outline our ideas for future work.

2 Related Work

We summarize related work in three sections: we describe some of the important work in sentiment analysis to give the general overview; followed by work on adaptation of a domain-independent polarity lexicon; and finally work that use a sentence-based or similar approach.

Research in sentiment analysis has started in the last 10–12 years, with increasing academic and commercial interest to the field. An elaborate survey of the previous works for sentiment analysis has been presented in [3] while we only summarize some important trends here.

In the earlier works, the document is typically viewed as a *bag of words* and its sentiment polarity is estimated from the average polarity of the words inside the document [5, 13–15]. Since looking at the whole document only as a bag of words is very simplistic, later work focused on analysis of phrases and sentences. Among these, some focused on subjectivity analysis of phrases/sentences, so as to make use of this information while determining the subjectivity of the document. In one of the early studies, Wiebe discovered subjective adjectives from corpora [16]. Then, Hatzivassiloglou and Wiebe [16] investigated the impacts of adjective orientation and gradability on sentence subjectivity. The goal behind this approach was to determine whether a given sentence is subjective or not, by examining the adjectives in that sentence. Subsequently, several studies focused on sentence-level or sub-sentence-level subjectivity detection in different domains. Some recent work also examined relations between word sense disambiguation and subjectivity, in order to extract sufficient information for a more accurate sentiment classification [18]. Wiebe et al.

introduces a broad survey of subjectivity recognition using various features and clues [19].

Determining the sentiment strength or polarity value of a given document, rather than simply classifying it as positive or negative, is a *regression* problem that is adressed using slightly different supervised learning techniques. In a regression problem, the task is to learn the mapping $y = f(x)$, where $x \in R^d$ and $y \in R$. It can be said that the regression problem is more difficult than the classification problem, as the latter can be accomplished once sentiment polarities are estimated. If one considers three sentiment categories (negative, neutral and positive), then treating the problem as a regression problem rather than a classification problem, may be the more appropriate approach since class labels are ordinal.

As the number of classes increase, the classification task becomes more difficult. For instance, classifying a review as positive or negative (two-class classification) is much easier than classifying it as very negative, negative, neutral, positive, and very positive (five-class problem). The problem is even more difficult when one considers objective as a separate category (e.g. negative, neutral, positive and objective), since objective and other (often neutral) classes may carry similar sentiment values. Here, the reader should note that opinionated text can be neutral ("The hotel was so so") and objective text can carry a sentiment value ("The hotel lacks a pool"). In approaching this problem, determining the sentiment subjectivity and sentiment strength can be done in two-steps.

A *polarity lexicon* indicates the sentiment polarity of words or phrases. Senti-Wordnet [4] and SenticNet [9] are two of the most commonly used polarity lexicons, for sentiment analysis. In [20], authors discuss three main approaches for opinion lexicon building: manual approach, dictionary-based approach, and corpus-based approach. The major shortcoming of the manual approach (e.g. [21]) is the cost (time and effort) to hand select words to build such a lexicon. There is also the possibility of missing important words that could be captured with automatic methods. Dictionary-based approaches (e.g. [4, 13, 22]) work by expanding a small set of seed opinion words, with the use of a lexical resource such as the WordNet [23]. Note that with these approaches, the resulting lexicons are domain-independent.

Corpus-based approaches can be used to learn domain-specific lexicons using a domain corpus of labeled reviews. Wilson et al. stress the importance of contextual polarity to differentiate from the prior polarity of a word [24]. They extract contextual polarities by defining several contextual features. In [25], a double propagation method is used to extract both sentiment words and features, combined with a polarity assignment method starting with a seed set of words. In [26], authors use linear programming to update the polarity of words based on specified hard or soft constraints. Another application of linear programming appears in [27] to learn a sentiment lexicon which is not only domain specific but also aspect-dependent. Another recent work expands a given dictionary of words with known polarities by first producing a new set of synonyms with polarities and using these to further deduce the polarities of other words [28]. Finally, a simple corpus-based domain adaptation technique proposed by Demiroz et al. is used in our system [1]. In this work, authors consider the tf-idf [29] scores of each word in positive and negative review sets, and adapt word polarities according to this difference.

The idea of sentence-level analysis is not new. Some researchers approached the problem by first finding subjective sentences in a review, with the hope of eliminating irrelevant sentences that would generate noise in terms of polarity estimation [8, 30]. Yet another approach is to exploit the structure in sentences, rather than seeing a review as a bag of words [13–15]. For instance in [13], conjunctions were analyzed to obtain the polarities of the words that are connected with the conjunct. In addition, Wilson et al. [32] raise the question of obtaining clause-level opinion strength as a preparation step for sentence-level sentiment analysis. In [33, 34] researchers focused on sentence polarities separately, again to obtain sentence polarities more correctly, with the goal of improving review polarity in turn. The first line polarity has also been used as a feature by [8].

Our work is motivated by our observation that the first and last lines of a review are often very indicative of the review polarity. Starting from this simple observation, we formulated more sophisticated features for sentence level sentiment analysis. In order to do that, we performed an in-depth analysis of different sentence types.

Our approach described in the remainining sections has two main parts: domain-adaptation of a general purpose polarity lexicon and sentiment analysis using the adapted lexicon and new, sentence-based features. We explain these two parts in Sects. 3.2 and 4, respectively. For domain-adaptation of a general purpose lexicon, we propose several variations of a simple method which is based on the delta tf-idf concept [35]. We have previously shown the benefits of using the adaptation technique independently [1], by using a simple sentiment analysis algorithm with and without domain adaptation of the used lexicon. We use the adapted polarity lexicon for feature extraction. For evaluating the sentiment of a given text, we propose some new and sentence-based features, based on the word polarities obtained from the adapted lexicon. Our state-of-the-art results on estimating overall document sentiment in two different domains, reported in Sect. 6 show the effectiveness of the proposed method.

3 Domain-Adaptation of a Polarity Lexicon

3.1 SentiWordNet

The polarity lexicon we use as the domain-independent lexicon is the SentiWordNet that consists of a list of words with their POS tags and three associated polarity scores $\langle pol^-, pol^=, pol^+ \rangle$ for each word [6]. The polarity scores indicate the measure of negativity, objectivity and positivity, and they sum up to 1. Some sample scores are provided in Table 1 from SentiWordNet.

As many other researchers have done, we simply select the dominant polarity of a word as its polarity and use the sign to indicate the polarity direction. The dominant polarity of a word w, denoted by $pol(w)$, is calculated as:

Table 1 Sample entries from SentiWordNet

Word	Type	Negative	Objective	Positive
Sufficient	JJ	0.75	0.125	0.125
Comfy	JJ	0.75	0.25	0.0
Moldy	JJ	0.375	0.625	0.0
Joke	NN	0.19	0.28	0.53
Fireplace	NN	0.0	1.0	0.0
Failed	VBD	0.28	0.72	0.0

$$pol(w) = \begin{cases} 0 & \text{if } max(pol^=, pol^+, pol^-) = pol^= \\ pol^+ & \text{else if } pol^+ \geq pol^- \\ -pol^- & \text{otherwise} \end{cases} \quad (1)$$

In other words, given the polarity triplet $\langle pol^-, pol^=, pol^+ \rangle$ for a word w, if the objective polarity is the maximum of the polarity scores, then the dominant polarity is 0. Otherwise, the dominant polarity is the maximum of the positive and negative polarity scores where pol^- becomes $-pol^-$ in the average polarity calculation. For example, the polarity triplet of the word "sufficient" is $\langle 0.75, 0.125, 0.125 \rangle$ $pol(\text{"sufficient"}) = -0.75$. Similarly, the polarity triplet of the word "moldy" is $\langle 0.375, 0.625, 0.0 \rangle$; hence $pol(\text{"moldy"}) = 0$.

An alternative way for calculating dominant polarity could be to completely ignore the objective polarity $pol^=$ and determine the $pol(w_i)$ of the word to be the maximum of pol^- and pol^+. With this method, the dominant polarity of the word "moldy" would be -0.375 instead of 0. However, we preferred the first approach as more appropriate, since many words appear as objective or dominantly objective in SentiWordNet.

3.2 Adapting a Domain-Independent Lexicon

The basic idea for domain adaptation is to learn the domain-specific polarities from labeled reviews in a given domain. For domain adaptation, we use the technique proposed in [1] with their best reported update mechanism. The proposed approach allows us to adapt a domain-independent lexicon such as SentiWordNet for a specific domain, by updating the polarities of only a small subset of the words. It was shown in [1] that updating the polarities of even a small set of words has a significant contribution to sentiment analysis accuracy.

This method analyzes the occurrence of the words in the lexicon in positive and negative reviews in a given domain. If a particular word occurs significantly more often in positive reviews than in negative reviews, then it is assumed that this word should have positive polarity for this domain, and vice versa. In this case, the polarity of that word is updated in the domain-specific lexicon.

While any domain-independent polarity lexicon can be used, we have adapted a commonly used lexicon, namely SentiWordNet [6]. Results with bigger and better lexicons such as SenticNet [9] are expected to be similar, albeit possibly showing smaller benefits.

In this method, inspired by [35], the tf-idf (term frequency-inverse document frequency) scores of each word is calculated separately for positive and negative review classes. The $tf(w, c)$ counts the occurrence of word w in class c, while $idf(w)$ is the proportion of documents where the word w occurs, discounting very frequently occurring words in the whole database (e.g. 'not', 'be') [36]. There are quite a few variants of tf-idf computations [29], and the tf-idf variant used by Demiroz et al. [1] is computed as:

$$tf.idf(w_i, +) = tf(w_i, +) \times idf(w_i) = log_e(tf(w_i, +) + 1) \times log_e(N/df(w_i))$$
$$tf.idf(w_i, -) = tf(w_i, -) \times idf(w_i) = log_e(tf(w_i, -) + 1) \times log_e(N/df(w_i))$$
$$(2)$$

where the first term to the right of the equality is the scaled term frequency (tf) and the second term is the scaled inverse document frequency (idf). The term $df(w_i)$ indicates the document frequency which is the number of documents in which w_i occurs and N is the total number of documents (reviews in our case) in the database. Then, the term $(\Delta tf)idf$ is defined as:

$$(\Delta tf)idf(w_i) = tf.idf(w_i, +) - tf.idf(w_i, -) \qquad (3)$$

Demiroz et al. [1] considers different alternatives about which polarities to update (e.g. the ones for which the $(\Delta tf)idf$ magnitude is large) and how to update them (e.g. use the $(\Delta tf)idf$ value as polarity or shift the original polarity value) and show that updating even a small percentage of all words in a lexicon improves sentiment analysis.

For adapting the domain-specific lexicon, we use the same update algorithm along with the best update method found in this work [1]. In this update method, we shift the polarities of the words that have the largest $(\Delta tf)idf$ scores in terms of absolute values. Hence, we consider both the largest and smallest $(\Delta tf)idf$ scores, suggesting the words with positive and negative connotations respectively. Once we select which words to adapt, we shift the original polarity values of those words towards their $(\Delta tf)idf$ scores by 0.4.

Table 2 Summary of features

Group name	Feature	Name
Basic	F_1	Average review polarity
	F_2	Review purity
	F_3	Review subjectivity
$(\Delta tf)idf$	F_4	Total $(\Delta tf)idf$ scores of all words
	F_5	Average review polarity, weighted by $(\Delta tf)idf$ scores
Seed words statistics	F_6	Freq. of seed words
	F_7	Avg. polarity of seed words
	F_8	Stdev. of polarities of seed words
Punctuation	F_9	# of Exclamation marks
	F_{10}	# of Question marks
	F_{11}	Number of positive smileys
	F_{12}	Number of negative smileys
Sentence-level	F_{13}	Avg. first line polarity
	F_{14}	Avg. last line polarity
	F_{15}	First line purity
	F_{16}	Last line purity
	F_{17}	Total $(\Delta tf)idf$ scores of words in the first line
	F_{18}	$(\Delta tf)idf$ weighted polarity of first line
	F_{19}	Total $(\Delta tf)idf$ scores of words in the last line
	F_{20}	$(\Delta tf)idf$ weighted polarity of last line
	F_{21}	Number of sentences in review
	F_{22}	Avg. pol. of subj. sentences
	F_{23}	Avg. pol. of pure sentences
	F_{24}	Avg. pol. of non-irrealis sentences

4 Sentence Based Sentiment Analysis Tool

For sentiment analysis of a given document or review, we propose and evaluate new features to be used in a word polarity-based approach to sentiment classification. The 24 features can be grouped in five listed in Table 2: (1) basic features, (2) $(\Delta tf)idf$ weighting of word polarities, (3) features based on the seed words statistics, (4) punctuation, and (5) sentence-level features.

Our approach depends on the existence of a sentiment lexicon that provides information about the semantic orientation of single or multiple terms. Specifically, we use the SentiWordNet [6] as the base lexicon and its domain-adapted version for domain-specific lexicon.

In the following sections, we define a review R as a sequence of sentences $R = S_1 S_2 S_3, \ldots, S_M$ where M is the number of sentences in R. The review R is also viewed as a sequence of words $w_1, \ldots, w_T$, where T is the total number of words in the review.

4.1 Basic Features

As the main features, we use review polarity, purity and subjectivity, which are commonly used in sentiment analysis. In our formulation $pol(w_j)$ denotes the dominant polarity of w_j of R, as obtained from SentiWordNet, and $|pol(w_j)|$ denotes the absolute polarity of w_j. We only include words with POS tags containing "NN", "JJ", "RB", and "VB" since these are the words that are possible sentiment-baring words in a review.

$$\text{Average review polarity} = \frac{1}{T} \sum_{j=1..T} pol(w_j) \qquad (4)$$

$$\text{Review purity}(\sum_{j=1..T} pol(w_j))/(\sum_{j=1..T} |pol(w_j)|) \qquad (5)$$

The review subjectivity is a binary variable that is 1 if one of the sentences in the review is deemed as subjective, as defined in Sect. 4.

4.1.1 $\Delta tf^* idf$ Features

We compute the $(\Delta tf)idf$ scores of the words in SentiWordNet from a training corpus in the given domain, in order to capture domain specificity as explained in Sect. 3.2. If the $(\Delta tf)idf$ score is positive, it indicates that a word is more associated with the positive class and vice versa, if negative. We computed these scores on the training set which is balanced in the number of positive and negative reviews.

We then extract two features using the $(\Delta tf)idf$ scores. In feature F_4, we compute the sum of the $(\Delta tf)idf$ scores of the unique words in a review. We expect that this feature may replace or complement the average review polarity obtained from the domain-independent lexicon. Note that the average $(\Delta tf)idf$ score of the words in the review would be very similar to the average polarity of the words (F_1) in the review; hence we preferred to use the sum even though it is dependent on the review length. As another feature F_5, we tried combining the two sources of information, where we weighted the polarities of all words in the review by their $(\Delta tf)idf$ scores (F_5).

Table 3 Chosen seed words

Positive word	Type	Negative word	Type
Great	JJ	Room	NN
Excellent	JJ	Desk	NN
Wonderful	JJ	Never	RB
Perfect	JJ	Worst	JJS
Fantastic	JJ	Manager	NN
Comfortable	JJ	Bad	JJ
Helpful	JJ	Night	NN
Friendly	JJ	Even	RB
Location	NN	Terrible	JJ
Lovely	JJ	Rude	JJ

4.1.2 Seed Word Statistics

Like some other researchers, we also use a smaller subset of the lexicon consisting of 20 clearly positive and 20 clearly negative seed words for the given domain, with the hope that they may indicate the reviews polarity with more certainty. To appreciate this approach, one can note that while a negative sentence might contain the word "good" ("the food was not good"), it is less likely for a negative sentence to contain the word "excellent" (e.g. "the food was not excellent"). In general, it is more likely to see a negative sentence containing a positive term, than a negative sentence containing a clearly positive seed word.

While we have computed the seed words automatically by analyzing the $(\Delta tf)idf$ scores of words, we assume that forming such a small list manually is feasible for any domain. To determine the seed words in the given domain, we first compute the $(\Delta tf)idf$ scores of all unique words in the corpus. Then, we sort these words by using their $(\Delta tf)idf$ scores and selected the top-20 positive and top-20 negative words in the list. These words then form the seed word set, called $SeedW$. We include a sub-sample of 10 positive and 10 negative seed words in Table 3.

We then define $SeedW(R)$ as the set of seed words that appear in review R and extract three features related to seed words in the review (features $F6 - F8$):

$$\text{Freq. of seed words} = |SeedW(R)|/|R| \tag{6}$$

$$\text{Avg. polarity of seed words} = \frac{1}{|SeedW(R)|} \sum_{w_j \in SeedW(R)} pol(w_j) \tag{7}$$

We also include the standard deviation of the polarity of seed words in the review, to capture if there are any disagreements.

4.1.3 Punctuation Features

We have four features related to punctuation. Two of these features were suggested before; namely, the number of exclamation marks and the number of question marks [37]. Note that an exclamation mark typically makes the stated emotion stronger ("the food was good!!"), but it can also be used to indicate an incredulous reaction (e.g. "the room did not have a window!"). On the other hand, the question mark can be used to detect objective/neutral sentences that may be otherwise classified as having sentimental polarity ("are the rooms big?").

Emoticons, in our case positive and negative smiley-faces, are also important sentiment-bearing symbols, as proposed in [38, 39]. As with the exclamation and question marks, smiley-faces may also have distinct context-specific meanings. For instance the positive smiley may be used positively, to indicate happiness (e.g. "the room had a view :)") or to make fun of something or agree with a joke.

Despite the ambiguities, we have included the two punctuation features and the two smiley-faces in our feature set, with the hope that statistics related to their usage may give some additional information to the classifier.

4.1.4 Sentence-Level Features

Often the first or last line in a review summarizes the overall review sentiment. This is certainly true for long reviews found in hotel reviews. For instance a title or first line such as "Excellent hotel!" clearly denotes the overall sentiment, no matter what is said in the details of the review. In this work, we propose to consider the review as a set of sentences and estimate the review sentiment by considering the types and sentiment strength of the constituent sentences.

Sentence-level features are extracted from (i) sentences in certain locations in the review (e.g. the first and last lines of the review) or (ii) certain types of sentences (e.g. subjective sentences). In particular, we consider subjective, pure and non-irrealis sentences and use features extracted from such sentences for detecting the review sentiment.

There are many possibilities in a sentence-based analysis. For instance, one can (i) consider only subjective sentences or (ii) use the features of subjective sentences as additional features in the system. We explored both of these approaches and then decided to add sentence-level features to the system.

In order to identify subjective sentences, we looked at if a sentence contains at least one subjective word or a smiley; if so that sentence is deemed as subjective. For subjectivity of the word, we adopted the same idea that was proposed in [40].

Similarly, we consider a sentence S_i as *pure* if its purity is greater than a fixed threshold τ. Sentence purity can be calculated as in Eq. 5, using only the words in the sentence. We experimented with different values of τ and for evaluation we used $\tau = 0.8$.

Table 4 Sentence-level features for a review R

F_{13}	Avg. first line polarity	$\frac{1}{|S_1|} \sum_{w \in S_1} pol(w)$
F_{14}	Avg. last line polarity	$\frac{1}{|S_M|} \sum_{w \in S_M} pol(w)$
F_{15}	First line purity	$[\sum_{w \in S_1} pol(w)]/[\sum_{w \in S_1} |pol(w)|]$
F_{16}	Last line purity	$[\sum_{w \in S_M} pol(w)]/[\sum_{w \in S_M} |pol(w)|]$
F_{17}	$(\Delta tf)idf$ weighted polarity of 1st line	$(\sum_{w \in S_1} \Delta tf * idf(w)) \times pol(w)$
F_{18}	Total $(\Delta tf)idf$ scores of 1st line	$\sum_{w \in S_1} \Delta tf * idf(w)$
F_{19}	$(\Delta tf)idf$ weighted polarity of last line	$(\sum_{w \in S_M} \Delta tf * idf(w)) \times pol(w)$
F_{20}	Total $(\Delta tf)idf$ scores of last line	$\sum_{w \in S_M} \Delta tf * idf(w)$
F_{21}	Number of sentences in review	M
F_{22}	Avg. pol. of subj. sentences	$\frac{1}{|subjS(R)|} \sum_{w \in subjW(R)} pol(w)$
F_{23}	Avg. pol. of pure sentences	$\frac{1}{|pureS(R)|} \sum_{w \in pure(R)} pol(w)$
F_{24}	Avg. pol. of non-irrealis sentences	$\frac{1}{|nonIrS(R)|} \sum_{w \in nonIr(R)} pol(w)$

We also looked at sentences containing *irrealis* terms, as they indicate the opposite sentiment than the sentiment carried by the constituent words (e.g. "I thought the hotel would have nicer rooms."). In order to determine irrealis sentences, the existence of the modal verbs "would", "could", or "should" is checked. If one of these modal verbs appears in the sentence, then these sentences are labeled as irrealis sentences, as was the case in [7]. Then, we chose *non-irrealis* sentences as our third sentence type for analysis.

These three sets of sentences in a review R are called $subS(R)$, $pureS(R)$ and $nonIrS(R)$. The sentence-based features ($F_{13} - F_{24}$) are given in Table 4:

We tried three different approaches for this purpose.

- In the first approach, each review is pruned to keep only the sentences that are possibly more useful (e.g. only subjective sentences) for sentiment analysis. For pruning, thresholds were set separately for each sentence-level feature. Sentences with absolute purity of at least 0.8 are defined as pure sentences. Pruning sentences in this way resulted in lower accuracy in general, due to loss of information.
- In the second approach, the polarities in special sentences (pure, subjective or no irrealis) are given higher weights while computing the average review polarity. In effect, this is a soft version of pruning, as the other sentences are given lower weight, rather than a weight of zero.
- In the third approach, we used the information extracted from sentence-level analysis as additional features (e.g. average polarity of subjective sentences were added as a new feature). This approach gave the best results and is used in the final system.

5 Classification

Given a review, we first apply feature extraction and represent the review by its features given in Table 2. We used a supervised learning approach to train a classifier to learn to classify a review into different sentiment classes or to assign a sentiment strength to it.

For two-class classification, we took 1- and 2-star reviews in the training set as negative samples and 4- and 5-star reviews as positive samples. For three-class classification, we instead trained a regression engine to estimate the sentiment strength and then decided on two thresholds delineating the negative-neutral and neutral-positive boundaries.

In the classification problem (2- to 5-class classification problems are considered for the hotel domain in literature), the performance measure is the classification accuracy; in other words, what percentage of queried reviews got classified correctly. In the case of regression, a natural error measure is the Mean Squared Error (MSE) or Mean Absolute Error (MAE), in other words the average squared or absolute error between the estimated strength and the ground truth. When using regression as a first step for classification, one can measure classification accuracy by using thresholds after estimating the sentiment strength.

As classifier, we used the Support Vector Machines (SVM) [41]. For the implementation, we used the LibSVM [43]. package in WEKA [44] for both train and test phases. The SVM requires two main parameters for training: the kernel and the cost (C) parameter. The kernel, cost and gamma parameters required for one of the kernels were decided on the validation set, using WEKA. For kernel, we tried the RBF & linear kernels and observed that the RBF kernel worked better than the linear kernel for our task.

For two-class classification, we used C-SVC (classification), RBF kernel and the parameter pair (10.0, 10.0). For regression, we used epsilon-SVR (regression) as SVM type and set the normalization to true by default. The cost and gamma parameters were the same as for classification, even though parameter optimization was done separately for this problem.

6 Experimental Evaluation

In this section, we provide an evaluation of the sentiment analysis engine. Our evaluation procedure is composed of two main parts. First, we report the effectiveness of different sets of features using star-rated reviews from the TripAdvisor website [2]. Next, we evaluate our overall system with state-of-the-art approaches on the hotel reviews dataset presented in [12].

6.1 Dataset

The TripAdvisor dataset consists of around 240,000 customer-supplied reviews of 1,850 hotels and was introduced by [42]. Each review is associated with a hotel and a star-rating, 1-star (most negative) to 5-star (most positive), chosen by the customer to indicate his/her evaluation.

For feature and overall system evaluation, we used a publicly available dataset that was collected from this corpus [43], in order to make the system comparable to the state-of-the-art approaches. This dataset contains around 90,000 hotel reviews, in three subsets: the train, validation and test subsets contain approximately 76,000, 6,000 and 13,000 reviews respectively. Each of these three subsets contains a balanced number of negative (1-star and 2-star) and positive (4-star and 5-star) reviews.

The dataset also includes neutral reviews (e.g. with a rating value of 3) that are used in three-class classification. For binary classification, these neutral reviews are omitted from the dataset.

For feature evaluation task, we first used the validation subset to select the best feature subsets using two appropriate classifiers on WEKA [44]. The validation dataset is also used to find the best parameters for the corresponding classifiers. Then, the test dataset is used to evaluate feature subsets and overall performance of the system, with the two selected classifiers.

6.2 Implementation

We computed $\Delta tf * idf$ scores of the words which have POSTags of noun, adjective, verb and adverb in the training set. Subsequently, we updated the dominant polarities of the words obtained from SentiWordNet [45] according to the polarity adaptation procedure explained in Sect. 3.2.

We calculated features as explained in Sect. 4 and generated intermediate files that represent a review as a set of features, along with its label. These intermediate files for the three sub-datasets (e.g. train, validation and test) were created by a Java implementation on Eclipse environment and given to WEKA [44].

For classification, we trained a Support Vector Machine classifier with the Sequential Minimal Optimization (SMO) algorithm, using the intermediate files as training data. For this, we used the LibSVM library which is included in WEKA environment [44]. Firstly, we observed that the RBF kernel worked better than other kernels for our purpose. Then, we found the best parameter pair for the cost and gamma parameters of this kernel and evaluated our overall system with these optimal paramters on WEKA [44].

For the purpose of various feature subsets evaluation, we used two different classifiers (SMO and Logistic) that are also integrated into WEKA environment [44] after we tried several other classifiers.

Table 5 The effect of feature subsets on two-class classification using the TripAdvisor Dataset [12]

Feature Subset	Accuracy (SMO) (%)	Accuracy (Logistic) (%)
Basic(F1-F3)	59.62	59.66
... + $(\Delta tf)idf$(F4-F5)	59.97	59.48
... + Seed Words(F6-F8)	59.97	59.48
... + Punctuation(F9-F12)	60.47	60.18
... + First&LastLine Avg. pol. and Purity(F13-F16)	60.60	60.62
... + First&LastLine $(\Delta tf)idf$(F17-F20)	60.74	60.67
... + Sentence count(F21)	60.70	60.78
... + Subj. Sentence Avg. pol.(F22)	63.76	64.27
... + Pure Sentence Avg. pol.(F23)	63.21	62.89
... + Non-Irrealis Sentence Avg. pol.(F24)	**63.76**	**64.27**
Basic + Seed Words(F6-F8)	59.62	59.66
Basic + Punctuation(F9-F12)	60.11	60.03
Basic + First&LastLine Avg. pol. and Purity(F13-F16)	59.97	59.94
Basic + First&LastLine $(\Delta tf)idf$(F17-F20)	60.28	59.72
Basic + Sentence count(F21)	60.05	59.93
Basic + Subj. Sentence Avg. pol.(F22)	61.27	60.27
Basic + Pure Sentence Avg. pol.(F23)	60.19	60.02
Basic + Non-Irrealis Sentence Avg. pol.(F24)	62.47	62.64

6.3 Results

6.3.1 Contributions of Feature Subsets on Overall Accuracy

The accuracies obtained on the two-class problem are given in Table 5 where there are two groups of results. In the upper half of the table, we provide the results as more features are incrementally added in the order listed in Table 2. Note that in this way, features that are added first have more of a chance to improve on the baseline accuracy, so in the lower half, we provide accuracy results when features are added one by one to the basic features.

When considering these results, we noted that most of the accuracy gains were obtained with punctuation features (59.97 to 60.47 % using SMO); the addition of subjective sentences (60.70 to 63.76 % using SMO); and the addition of non-irrealis features (63.21 to 63.76 % using SMO).

Also, we noted that there is no improvement when features related to seed words statistics are added on top of the basic plus $(\Delta tf)idf$ features. This shows that seed words related features do not bring extra information to the system. Although, seed words seem to have no effect on overall accuracy, we still included seed words statistics features for the sake of completeness.

6.3.2 Overall Engine Comparison with Previous Systems

We provide state-of-the-art sentiment analysis performance results obtained using reviews from the TripAdvisor website in Table 6. Unfortunately not all systems report results that are directly comparable: they may differ in the tested data set; the reported accuracy or error measure; or the classification problem. In Table 6, different systems are grouped according to the number of classes. For instance some systems have reported their performance in the binary classification problem of separating 1- or 2-star reviews, from the 4- or 5-star reviews.

As one can see, the best results so far are obtained by Bespalov et al. using the LDA approach, with 6.90 % error rate on the binary classification problem [11, 43], while our error rate for this task is 13.23 %. In terms of the f-measure, our results surpass previously reported f-measures, with an f-measure of 0.87 for the binary classification problem and 0.64 for the three-class problem.

For the 5-class classification task, the best results achieved so far are again by Bespalov et al. with 40.76 % error rate [11, 43]. We handled the 5-class classification task as a regression problem and obtained the regression values for class labels of reviews from WEKA [44]. This gave a Mean Absolute Error of 0.43 on the test set. This can be seen as a review with +1 target value being assigned a sentiment strength of 0.57 $(1-0.43)$. When we rounded the estimated regression values (e.g. 1.8 became 2 while 1.3 became 1) and obtained classification in this way, the misclassification error is measured to be 56.25 %. While this rate is high, it actually highlights the difficulty of the 5-class classification problem. Note that a random classifier would be expected to be accurate for about one in five cases, or have an error rate of 80 %.

6.3.3 Discussion

As can be seen in Table 6, our system with the newly proposed features provides one of the best results obtained so far, except for the work presented in [11, 12].

It is noteworthy to mention that [12] is a more recent version of [11] and they both use LDA as the core approach. Topic models learned by methods such as LDA requires re-training when a new topic comes. In contrast, our system uses word polarities; therefore it is very simple and fast.

Table 6 State of the art results on the TripAdvisor Corpus

Previous work	Dataset	F-measure	Error rate	Task
Peter et al. [46]	103000	0.83	–	Binary: 1 versus {4,5}
Gindl et al. [47]	1800	0.79	–	Binary: {1,2} versus {4,5}
Gezici et al. [48]	6000	0.81	–	Binary: {1,2} versus {4,5}
Bespalov et al. [11]	96000[a]	–	7.37	Binary: {1,2} versus {4,5}
Bespalov et al. [12]	96000	–	6.90	Binary: {1,2} versus {4,5}
This work	96000	0.87	13.23	Binary: {1,2} versus {4,5}
Grabner et al. [49]	1000	0.55	–	Three-class: {1,2}, {3}, {4,5}
This work	96000	0.64	36.50	Three-class: {1,2}, {3}, {4,5}
Bespalov et al. [11]	96000[a]	–	49.20	Five-class
Bespalov et al. [12]	96000	–	40.76	Five-class
This work	96000	–	56.25	Five-class

[a]This dataset is different than the dataset released by Bespalov et al. [12]; so the results are not directly comparable

7 Conclusions and Future Work

We tried to bridge the gap between word-level polarities and review-level polarity through an intermediate step of sentence-level analysis of the reviews. We formulated new features for sentence-level sentiment analysis by an in-depth analysis of the sentences.

We implemented the proposed features and evaluated them on a publicly available dataset of TripAdvisor reviews [12], to show the effect of sentence-level features on polarity classification. We observed that sentence-level features indeed have an effect on sentiment classification accuracy; therefore, we conclude that sentences do matter in sentiment analysis and they may be even more useful in more diverse datasets such as blogs.

We also evaluated our domain-adapted engine on the same dataset of TripAdvisor hotel reviews and summarized state-of-the-art results in that domain. The variability of the datasets and accuracy measures make the reported results difficult to compare directly. Nonetheless, one can observe that two-class classification of text into positive and negative classes can be done quite robustly, while the five-class classification (required for assigning a star-rating) requires more work.

As future work, we will consider using word embeddings that have been shown to be successful in different problems [47], along with our existing approach. Sentence-based analysis can also be explored further to identify essential sentences in a review or for highlighting important sentences for review summarization.

Acknowledgments This work was partially funded by European Commission, FP7, under UBIPOL (Ubiquitous Participation Platform for Policy Making) Project (www.ubipol.eu). Dr. Dilek Tapucu was a post-doctoral researcher at Sabanci University at the time of this project.

References

1. Demiroz, G., Yanikoglu, B. Tapucu, D., Saygin, Y.: Learning domain-specific polarity lexicons, In: 2012 IEEE 12th International Conference on Data Mining Workshops (ICDMW), pp. 674–679 (2012)
2. The TripAdvisor website. http://www.tripadvisor.com [TripAdvisor LLC]. Accessed in 2012
3. Pang, B., Lee, L.: Opinion mining and sentiment analysis. Found. Trends Inf. Retrieval 2(1–2), 1–135 (2008)
4. Turney, P.D.: Thumbs up or thumbs down?: semantic orientation applied to unsupervised classification of reviews. In: Proceedings of the 40th Annual Meeting on Association for Computational Linguistics, pp. 417–424. Association for Computational Linguistics (2002)
5. Pang, B., Lee, L., Vaithyanathan, S.: Thumbs up?: sentiment classification using machine learning techniques. In: Proceedings of the ACL-02 conference on Empirical methods in natural language processing, vol. 10, pp. 79–86. Association for Computational Linguistics (2002)
6. Esuli, A., Sebastiani, F.: SentiWordNet: a publicly available lexical resource for opinion mining. In: Proceedings of the 5th Conference on Language Resources and Evaluation (LREC06), pp. 417–422 (2006)
7. Taboada, M., Brooke, J., Tofiloski, M., Voll, K.D., Stede, M.: Lexicon-based methods for sentiment analysis. Comput. Linguist. 37(2), 267–307 (2011)
8. Zhao, J., Liu, K., Wang, G.: Adding redundant features for crfs-based sentence sentiment classification. In: Proceedings of the 2008 Conference on Empirical Methods in Natural Language Processing, pp. 117–126 (2008)
9. Poria, S., Gelbukh, A.F., Cambria, E., Das, D., Bandyopadhyay, S.: Enriching SenticNet polarity scores through semi-supervised fuzzy clustering. In: Vreeken, J., Ling, C., Zaki, M.J., Siebes, A., Yu, J.X., Goethals, B., Webb, G.I., Wu, X. (eds.) ICDM Workshops, pp. 709–716. IEEE Computer Society (2012)
10. Yu, H., Hatzivassiloglou, V.: Towards answering opinion questions: separating facts from opinions and identifying the polarity of opinion sentences. In: Proceedings of the 2003 conference on Empirical methods in Natural Language Processing, pp. 129–136. Association for Computational Linguistics (2003)
11. Bespalov, D., Bai, B., Qi, Y., Shokoufandeh, A.: Sentiment classification based on supervised latent n-gram analysis. In: Proceedings of the 20th ACM International Conference on Information and Knowledge Management, pp. 375–382. ACM (2011)
12. Bespalov, D., Qi, Y., Bai, B., Shokoufandeh, A.: Sentiment lassification with supervised sequence embedding. In: Machine Learning and Knowledge Discovery in Databases, pp. 159–174. Springer (2012)
13. Hatzivassiloglou, V., Mckeown, K.R.: Predicting the semantic orientation of adjectives. In: Proceedings of ACL-97, 35th Annual Meeting of the Association for Computational Linguistics, pp. 174–181. Association for Computational Linguistics (1997)
14. Mao, Y., Lebanon, G.: Isotonic conditional random fields and local sentiment flow. Adv. Neural Inf. Process. Syst. 19, 961 (2007)
15. Pang, B., Lee, L.: A sentimental education: Sentiment analysis using subjectivity summarization based on minimum cuts. In: Proceedings of the 42nd annual meeting on Association for Computational Linguistics, p. 271. Association for Computational Linguistics (2004)
16. Wiebe, J.M.: Learning subjective adjectives from corpora. In: In AAAI, pp. 735–740 (2000)
17. Hatzivassiloglou, V., Wiebe, J.: Effects of adjective orientation and gradability on sentence subjectivity. In: Proceedings of the 18th Conference on Computational Linguistics, vol. 2, pp. 299–305. Universität des Saarlandes, Saarbrücken, Germany, July 31–Aug 4 (2000)
18. Wiebe, J., Mihalcea, R.: Word sense and subjectivity. In: Proceedings of the 21st International Conference on Computational Linguistics and the 44th annual meeting of the Association for Computational Linguistics, pp. 1065–1072. Association for Computational Linguistics (2006)
19. Wiebe, J., Wilson, T., Bruce, R., Bell, M., Martin, M.: Learning subjective language. Comput. Linguist. 30(3), 277–308 (2004)

20. Liu, B., Zhang, L.: A survey of opinion mining and sentiment analysis. In: Mining Text Data, pp. 415–463. Springer (2012)
21. Das, S.R., Chen, M.Y.: Yahoo! for amazon: sentiment extraction from small talk on the web. Manage. Sci. **53**(9), 1375–1388 (2007)
22. Turney, P.D., Littman, M.L.: Measuring praise and criticism: inference of semantic orientation from association. ACM Trans. Inf. Syst. (TOIS) **21**(4), 315–346 (2003)
23. Miller, G.A.: Wordnet: a lexical database for english. Commun. ACM **38**(11), 39–41 (1995)
24. Wilson, T., Wiebe, J., Hoffmann, P.: Recognizing contextual polarity: an exploration of features for phrase-level sentiment analysis. Comput. Linguist. **35**(3), 399–433 (2009)
25. Qiu, G., Liu, B., Bu, J., Chen, C.: Expanding domain sentiment lexicon through double propagation. In: Proceedings of the 21st international jont conference on Artifical intelligence, pp. 1199–1204 (2009)
26. Choi, Y., Cardie, C.: Adapting a polarity lexicon using integer linear programming for domain-specific sentiment classification. In: Proceedings of the 2009 Conference on Empirical Methods in Natural Language Processing, pp. 590–598 (2009)
27. Dragut, E.C., Yu, C., Sistla, P., Meng, W.: Construction of a sentimental word dictionary. In: Proceedings of the 19th ACM International Conference on Information and Knowledge Management, CIKM '10, pp. 1761–1764. ACM, New York, NY, USA (2010)
28. Lu, Y., Castellanos, M., Dayal, U., Zhai, C.: Automatic construction of a context-aware sentiment lexicon: an optimization approach. In: Proceedings of the 20th International Conference on World Wide Web, WWW '11, pp. 347–356. ACM, New York, NY, USA (2011)
29. Paltoglou, G., Gobron, S., Skowron, M., Thelwall, M., Thalmann, D.: Sentiment analysis of informal textual communication in cyberspace. Proc. Engage 13–25 (2010)
30. McDonald, R., Hannan, K., Neylon, T., Wells, M., Reynar, J.: Structured models for fine-to-coarse sentiment analysis. In: Annual Meeting-Association For Computational Linguistics, vol. 45, p. 432 (2007)
31. Kim, S.-M., Hovy, E.: Automatic detection of opinion bearing words and sentences. In: Proceedings of IJCNLP, vol. 5 (2005)
32. Wilson, T., Wiebe, J., Hoffmann, P.: Recognizing contextual polarity in phrase-level sentiment analysis. In: Proceedings of the Conference on Human Language Technology and Empirical Methods in Natural Language Processing, pp. 347–354. Association for Computational Linguistics (2005)
33. Meena, A., Prabhakar, T.: Sentence level sentiment analysis in the presence of conjuncts using linguistic analysis. In: Advances in Information Retrieval, pp. 573–580. Springer (2007)
34. Martineau, J., Finin, T.: Delta tfidf: an improved feature space for sentiment analysis. In: ICWSM (2009)
35. Salton, G., Wong, A., Yang, C.-S.: A vector space model for automatic indexing. Commun. ACM **18**(11), 613–620 (1975)
36. Denecke, K.: How to assess customer opinions beyond language barriers? In: ICDIM, IEEE, pp. 430–435 (2008)
37. Bifet, A., Frank, E.: Sentiment knowledge discovery in twitter streaming data. In: Discovery Science, pp. 1–15. Springer (2010)
38. Pak, A., Paroubek, P.: Twitter as a corpus for sentiment analysis and opinion mining. In: LREC (2010)
39. Zhang, E., Zhang, Y.: Ucsc on trec 2006 blog opinion mining. In: Text Retrieval Conference (2006)
40. Chang, C.-C., Lin, C.-J.: Libsvm: a library for support vector machines. ACM Trans. Intell. Syst. Technol. (TIST) **2**(3), 27 (2011)
41. Wang, H., Lu, Y., Zhai, C.: Latent aspect rating analysis on review text data: a rating regression approach. In: Proceedings of the 16th ACM SIGKDD International Conference on Knowledge Discovery and Data Mining, pp. 783–792 (2010)
42. Bespalov, D., Qi, Y., Bai, B., Shokoufandeh, A.: Sentiment classification with supervised sequence embedding. In: Flach,P.A. Bie, T.D., Cristianini, N. (eds.) ECML/PKDD (1). Lecture Notes in Computer Science, vol. 7523, pp. 159–174. Springer (2012)

43. Hall, M., Frank, E., Holmes, G., Pfahringer, B., Reutemann, P., Witten, I.H.: The weka data mining software: an update. ACM SIGKDD Explor. Newsl. **11**(1), 10–18 (2009)
44. Esuli, A., Sebastiani, F.: Determining term subjectivity and term orientation for opinion mining. In: Proceedings of EACL, vol. 6, pp. 193–200 (2006)
45. Lau, R.Y.K., Lai, C.L., Bruza, P.B., Wong, K.F.: Leveraging web 2.0 data for scalable semi-supervised learning of domain-specific sentiment lexicons. In: Proceedings of the 20th ACM International Conference on Information and Knowledge Management, CIKM '11, pp. 2457–2460. ACM, New York, NY, USA (2011)
46. Gindl, S., Weichselbraun, A., Scharl, A.: Cross-domain contextualisation of sentiment lexicons. In: Proceedings of the 19th European Conference on Artificial Intelligence (ECAI), 16 Aug 2010
47. Gezici, G., Yanikoglu, B., Tapucu, D., Saygın, Y.: New features for sentiment analysis: Do sentences matter?. In: SDAD 2012 The 1st International Workshop on Sentiment Discovery from Affective Data, p. 5 (2012)
48. Gräbner, D., Zanker, M., Fliedl, G., Fuchs, M.: Classification of customer reviews based on sentiment analysis. In: Information and Communication Technologies in Tourism 2012, pp. 460–470. Springer (2012)
49. Maas, A.L., Daly, R.E., Pham, P.T., Huang, D., Ng, A.Y., Potts, C.: Learning word vectors for sentiment analysis. In: Lin, D., Matsumoto, Y., Mihalcea, R. (eds.) ACL, pp. 142–150. The Association for Computer Linguistics (2011)

Entity-Based Opinion Mining from Text and Multimedia

Diana Maynard and Jonathon Hare

Abstract This paper describes the approach we take to the analysis of social media, combining opinion mining from text and multimedia (images, videos, etc.), and centred on entity and event recognition. We examine a particular use case, which is to help archivists select material for inclusion in an archive of social media for preserving community memories, moving towards structured preservation around semantic categories. The textual approach we take is rule-based and builds on a number of sub-components, taking into account issues inherent in social media such as noisy ungrammatical text, use of swear words, sarcasm etc. The analysis of multimedia content complements this work in order to help resolve ambiguity and to provide further contextual information. We provide two main innovations in this work: first, the novel combination of text and multimedia opinion mining tools; and second, the adaptation of NLP tools for opinion mining specific to the problems of social media.

1 Introduction

Social web analysis is all about the users who are actively engaged and generate content. This content is dynamic, reflecting the societal and sentimental fluctuations of the authors as well as the ever-changing use of language. Social networks are pools of a wide range of articulation methods, from simple "Like" buttons to complete articles, their content representing the diversity of opinions of the public. User activities on social networking sites are often triggered by specific events and related entities (e.g. sports events, celebrations, crises, news articles) and topics (e.g. global warming, financial crisis, swine flu).

D. Maynard (✉)
Department of Computer Science, University of Sheffield,
Regent Court, 211 Portobello, Sheffield S1 4DP, UK
e-mail: d.maynard@sheffield.ac.uk

J. Hare
Electronics and Computer Science, University of Southampton, Southampton,
Hampshire SO17 1BJ, UK
e-mail: jsh2@ecs.soton.ac.uk

© Springer International Publishing Switzerland 2015   65
M.M. Gaber et al. (eds.), *Advances in Social Media Analysis*,
Studies in Computational Intelligence 602,
DOI 10.1007/978-3-319-18458-6_4

With the rapidly growing volume of resources on the Web, archiving this material becomes an important challenge. The notion of community memories extends traditional Web archives with related data from a variety of sources. In order to include this information, a semantically-aware and socially-driven preservation model is a natural way to go: the exploitation of Web 2.0 and the wisdom of crowds can make web archiving a more selective and meaning-based process. The analysis of social media can help archivists select material for inclusion, while social media mining can enrich archives, moving towards structured preservation around semantic categories. The ARCOMEM project[1] aims to extract, analyse and correlate such information from a vast number of heterogeneous Web resources, including multimedia, based on an iterative cycle consisting of (1) targeted archiving/crawling of Web objects; (2) entity, topic, opinion and event (ETOE) extraction and (3) refinement of crawling strategy. In this paper, we focus on the challenges in the development of opinion mining tools from both textual and multimedia content. It focuses on two very different domains: socially aware federated political archiving (realised by the national parliaments of Greece and Austria), and socially contextualized broadcaster web archiving (realised by two large multimedia broadcasting organizations based in Germany: Sudwestrundfunk and Deutsche Welle). The aim is to help journalists and archivists answer questions such as what the opinions are on crucial social events, how they are distributed, how they have evolved, who the opinion leaders are, and what their impact and influence is.

Alongside natural language, a large number of the interactions which occur between social web participants include other media, in particular images. Determining whether a specific non-textual media item is performing as an opinion-forming device in some interaction becomes an important challenge, more so when the textual content of some interaction is small or has no strong sentiment. Attempting to determine a sentiment value for an image clearly presents great challenges, and this field of research is still in its infancy. We describe here some work we have been undertaking, firstly to attempt to provide a sentiment value from an image outside of any specific context, and secondly to utilise the multimodal nature of the social web to assist the sentiment analysis of either the multimedia or the text.

2 Related Work

While much work has recently focused on the analysis of social media in order to get a feel for what people think about current topics of interest, there are, however, still many challenges to be faced. State of the art opinion mining approaches that focus on product reviews and so on are not necessarily suitable for our task, partly because they typically operate within a single narrow domain, and partly because the target of the opinion is either known in advance or at least has a limited subset (e.g. film titles, product names, companies, political parties, etc.).

---

[1]http://www.arcomem.eu.

In general, sentiment detection techniques can be roughly divided into lexicon-based methods [1] and machine-learning methods, e.g. [2]. Lexicon-based methods rely on a sentiment lexicon, a collection of known and pre-compiled sentiment terms. Machine learning approaches make use of syntactic and/or linguistic features, and hybrid approaches are very common, with sentiment lexicons playing a key role in the majority of methods. For example, [3] establish the polarity of reviews by identifying the polarity of the adjectives that appear in them, with a reported accuracy of about 10 % higher than pure machine learning techniques. However, such relatively successful techniques often fail when moved to new domains or text types, because they are inflexible regarding the ambiguity of sentiment terms. The context in which a term is used can change its meaning, particularly for adjectives in sentiment lexicons [4]. Several evaluations have shown the usefulness of contextual information [5], and have identified context words with a high impact on the polarity of ambiguous terms [6]. A further bottleneck is the time-consuming creation of these sentiment dictionaries, though solutions have been proposed in the form of crowdsourcing techniques.[2]

Almost all the work on opinion mining from Twitter has used machine learning techniques. Pak and Paroubek [7] aimed to classify arbitrary tweets on the basis of positive, negative and neutral sentiment, constructing a simple binary classifier which used n-gram and POS features, and trained on instances which had been annotated according to the existence of positive and negative emoticons. Their approach has much in common with an earlier sentiment classifier constructed by [8], which also used unigrams, bigrams and POS tags, though the former demonstrated through analysis that the distribution of certain POS tags varies between positive and negative posts. One of the reasons for the relative paucity of linguistic techniques for opinion mining on social media is most likely due to the difficulties in using NLP on low quality text [9]; for example the Stanford NER drops from 90.8 % F1 to 45.88 % when applied to a corpus of tweets [10].

One feature of social media is the much higher use of sarcasm, irony and swear words, which we deal specifically with in this work (see Sect. 3.2). There have been a number of recent works attempting to detect sarcasm in tweets and other user-generated content [11–14], with accuracy typically around 70–80 %. These mostly train over a set of tweets with the #sarcasm and/or #irony hashtags, but all simply try to classify whether a sentence or tweet is sarcastic or not (and occasionally, into a set of pre-defined sarcasm types). However, none of these approaches go beyond the initial classification step and thus cannot predict how the sarcasm will affect the sentiment expressed. This is one of the issues that we tackle in this work.

Extracting sentiment from images is still a research area that is in its infancy and not yet prolifically published. However, those published often use small datasets for their ground truth on which to build SVM classifiers. Evaluations show systems often respond only a little better than chance for trained emotions from general images [15]. The implication is that the feature selection for such classification is difficult. [16] used a set of colour features for classifying their small ground-truth dataset, also

---

[2]http://apps.facebook.com/sentiment-quiz.

using SVMs, and publish an accuracy of around 87 %. In our work, we expand this colour-based approach to use other features and also use the wisdom of the crowd for selecting a large ground-truth dataset.

Other papers have begun to hint at the multimodal nature of web-based image sentiment. Earlier work, such as [17], is concerned with similar multimodal image annotation, but not specifically for sentiment. They use latent semantic spaces for correlating image features and text in a single feature space. In this paper, we describe the work we have been undertaking in using text and images together to form sentiment for social media.

3 Opinion Mining from Text

3.1 Challenges

There are many challenges inherent in applying typical opinion mining and sentiment analysis techniques to social media. Microposts such as tweets are, in some sense, the most challenging text type for text mining tools, and in particular for opinion mining, since the genre is noisy, documents have little context and assume much implicit knowledge, and utterances are often short. As such, conventional NLP tools typically do not perform well when faced with tweets [18], and their performance also negatively affects any following processing steps.

Ambiguity is a particular problem for tweets, since we cannot easily make use of co-reference information: unlike in blog posts and comments, tweets do not typically follow a conversation thread, and appear much more in isolation from other tweets. Here we can make use of the image and multimodal analysis: for example if Paris is mentioned and we have a picture of the Eiffel Tower, we can disambiguate the correct meaning of Paris to the city in France. Tweets also exhibit much more language variation, and make frequent use of emoticons, abbreviations and hashtags, which can form an important part of the meaning. Typically, they also contain extensive use of irony and sarcasm, which are particularly difficult for a machine to detect. On the other hand, their terseness can also be beneficial in focusing the topics more explicitly: it is very rare for a single tweet to be related to more than one topic, which can thus aid disambiguation by emphasising situational relatedness.

In longer posts such as blogs, comments on news articles and so on, a further challenge is raised by the tracking of changing and conflicting interpretations in discussion threads. We investigate first steps towards a consistent model allowing for the pinpointing of opinion holders and targets within a thread (leveraging the information on relevant entities extracted).

We refer the reader to [18] for our work on twitter-specific IE, which we use as pre-processing for the opinion mining described below. It is not just tweets that are problematic, however; sarcasm and noisy language from other social media forms also have an impact. In the following section, we demonstrate some ways in which we deal with this.

3.2 Opinion Mining Application

Our approach is a rule-based one similar to that used by [1], focusing on building up a number of sub-components which all have an effect on the score and polarity of a sentiment. In contrast, however, our opinion mining component finds opinions relating to previously identified entities and events in the text. The core opinion mining component is described in [19], so we shall only give an overview here, and focus on some issues specific to social media which were not dealt with in that work, such as sarcasm detection and hashtag decomposition.

For the pre-processing, we use the TwitIE pipeline [18], depicted in Fig. 1. This is a customisation of ANNIE, the general purpose IE recognition application in GATE, developed specifically to handle social media content, and which has been tested most extensively on microblog messages. In the picture, re-used ANNIE components are shown in blue (dashed) boxes, whereas the ones in red (dotted) boxes are new and specific to the microblog genre.

The detection of the opinions is performed via a number of different phases: detecting positive, negative and neutral words, identifying factual or opinionated versus questions or doubtful statements, identifying negatives, sarcasm and irony, analysing hashtags, and detecting extra-linguistic clues such as smileys. The application involves a set of grammars which create annotations on segments of text. The grammar rules use information from gazetteers combined with linguistic features

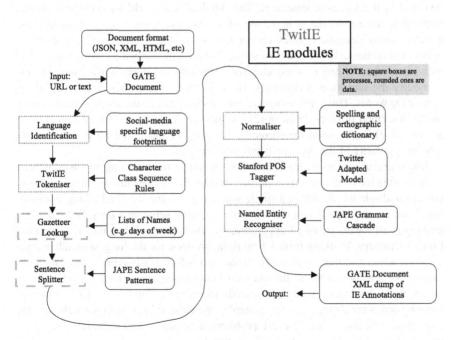

Fig. 1 The TwitIE information extraction pipeline

(POS tags etc.) and contextual information to build up a set of annotations and features, which can be modified at any time by further rules. The set of gazetteer lists contains useful clues and context words: for example, we have developed a gazetteer of affect/emotion words from WordNet [20]. The lists have been modified and extended manually to improve their quality.

Once sentiment words have been matched, we find a linguistic relation between these and an entity or event in the sentence or phrase. A Sentiment annotation is created for that entity or event, with features denoting the polarity (positive or negative) and the polarity score. Scores are based on the initial sentiment word score, and intensified or decreased by any modifiers such as swear words, adverbs, negation, sarcasm etc., as explained next.

Swear words are particularly prolific on Twitter, especially on topics such as popular culture, politics and religion, where people tend to have very strong views. To deal with these, we match against a gazetteer list of swear words and phrases, which was created manually from various lists found on the web and from manual inspection of the data, including some words acquired by collecting tweets with swear words as hashtags (since these also often contain more swear words in the main text of the tweet).

Sarcasm occurs frequently in user-generated content such as blogs, forums and microposts, especially in English, and is inherently difficult to analyse, not only for a machine but even for a human. One needs to have a good understanding of the context of the situation, the culture in question, and perhaps the very specific topic or people involved in the sarcastic statement. This kind of real-world knowledge is almost impossible for a machine to make use of. Furthermore, even correctly identifying a statement as sarcastic is often insufficient to be able to analyse it, especially in terms of sentiment, due to issues of scope. We have developed a component to deal with sarcasm according to some sarcastic keywords and in particular hashtags, and to modify the sentiment according to the effect the sarcasm has on the sentiment-containing words. This involves a set of rules to determining the scope of the sarcasm, and is described in more detail in [21]. Initial experiments on twitter data gave us a score of 91% Precision for correctly detecting sarcasm, and 80% for correctly detecting polarity of the opinion in sarcastic tweets.

Much useful sentiment information, particularly for detecting sarcasm, is contained within hashtags, but this is problematic to identify because hashtags typically contain multiple words within a single token, e.g. #notreally. If a hashtag is camel-cased, we use the capitalisation information to create separate tokens. Second, if the hashtag is all lowercase or all uppercase, we try to form a token match against the Linux dictionary. Working from left to right, we look for the longest match against a known word, and then continue from the next offset. If a combination of matches can be found without a break, the individual components are converted to tokens. In our example, #notreally would be correctly identified as "not" + "really". However, some hashtags are ambiguous: for example, "#greatstart" gets split wrongly into the two tokens "greats" + "tart". These problems are hard to deal with; in some cases, we could make use of contextual information to assist.

We conducted an experiment to measure the accuracy of hashtag decomposition, using a corpus of 1000 tweets randomly selected from the US elections crawl that we undertook in the project. 944 hashtags were detected in this corpus, of which 408 were identified as multiword hashtags (we included combinations of letters and numbers as multiword, but not abbreviations). 281 were camelcased and/or combinations of letters and nubers, 27 were foreign words, and the remaining 100 had no obvious token-distinguishing features. Evaluation on the hard-to-recognise cases (non-camel-cased multiword hashtags) produced scores of 86.91 % Precision, 90 % Recall, and an F-measure of 88.43 %. We conducted a second experiment using a gold standard set of tokenised hashtags extracted from a larger corpus of general tweets that we also annotated manually. This gold standard set contained 2010 hashtags and 4538 tokens. The system achieved 98.12 % Precision and 96.41 % Recall, and an F1 of 97.25 %.

For a fairly simple solution, these initial results are pleasing. One error is due to the presence of unknown named entities (people, locations and organisations) forming part of the hashtag. While some of these are recognised by our gazetteer lookup (especially locations), many of them are unknown. The named entity recognition component in GATE cannot identify these until they have been correctly tokenised, so we have a circular problem. We could also investigate using a language modelling approach based on unigram or bigram frequencies, such as that used by Berardi et al. [22].

In addition to using the sentiment information from these hashtags, we also collect new hashtags that typically indicate sarcasm, since often more than one sarcastic hashtag is used. For this, we used the GATE gazetteer list collector to collect pairs of hashtags where one was known to be sarcastic, and examined the second hashtag manually. From this we were able to identify a further set of sarcasm-indicating hashtags, such as #thanksdude, #yay etc. Further investigation needs to be performed on these to check how frequently they actually indicate sarcasm when used on their own, but preliminary analysis was promising.

Finally, emoticons are processed like other sentiment-bearing words, according to another gazetteer list, if they occur in combination with an entity or event. For example, the tweet "They all voted Tory :-(" would be annotated as negative with respect to the target "Tory". Otherwise, as for swear words, if a sentence contains a smiley but no other entity or event, the sentence gets annotated as sentiment-bearing, with the value of that of the smiley from the gazetteer list.

Once all the subcomponents have been run over the text, a final output is produced for each sentiment-bearing segment, with a polarity (positive or negative) and a score, based on combining the individual scores from the various components (for example, the negation component typically reverses the polarity, the adverbial component increases the strength of the sentiment, and so on). Aggregation of sentiment then takes place for all mentions of the same entity/event in a document, so that summaries can be created. This part is described in detail in [23].

Evaluation of the system has been performed on a variety of domain types, and has also been carried out for the individual subcomponents, although improvements to the system are still ongoing. The focus is on precision rather than recall in order to minimise wrongly opinionated results being generated, which would negatively affect the end system. On a dataset contining 70 manually annotated tweets about the US election, we achieved 87.50 % Precision and 63 % Recall for detecting opinion-ated tweets. For identifying the correct polarity of the opinions, we achieved 85.71 % Precision and 85.72 % Recall, which is very promising given the level of difficulty of the task.

4 Mining Images and Their Context

Images are often used to illustrate the opinions expressed by the text of a partic-ular media item. By themselves, images also have the ability to convey and elicit opinions, emotions and sentiments. In order to investigate how images are used in the opinion formation process, we have been developing tools that allow in-depth analysis of specific elements within an image to be used to quantify elements of opinion and sentiment, and allow the reuse of images within an archive or corpus to be contextualised with respect to diverse time and opinion axes.

4.1 Challenges

The main challenge with annotating non-textual media is that the underlying tokens within it are considerably less explicit than in textual media. In images and video, these underlying tokens are groups of pixels (compared with groups of characters [words] in text). As well as having multiple dimensions, the tokens have considerably more variation when representing exactly the same concept, and so using dictionaries and other traditional text-based techniques often becomes impractical. State of the art computer vision and automated image understanding is still a relatively immature subject for most general applications. This "semantic gap" between what computer vision can achieve and the level of understanding required for tasks such as sentiment analysis is why extracting opinions from images is so difficult.

Even though computer vision is challenging, considerable advances have been made in recent years. This is in particular true for the detection and recognition of certain types of objects or entities. In terms of the types of entities often recognised by Named Entity Recognition (NER) tasks in test documents, there are a number of relatively mature visual equivalents:

1. The detection of *Person* entities in images can be acheived in a fairly robust manner by detecting human faces, and face recognition technologies can help recognise and disambiguate the specific person. This is discussed in more detail in Sect. 4.2.
2. *Organisation* entities can be detected and recognised in images by looking for cer-tain indicators, such as the logo of the organisation. Techniques that can robustly

detect rigid patterns in images (such as logos) are common-place in modern computer vision (e.g. [24, 25]).

3. The recognition of *Place* entities is currently a hot topic in the multimedia analysis community, and a number of techniques for determining where an image was taken have been proposed. From a purely visual analysis point of view, these techniques tend to either work by directly matching the image against large datasets of images with known locations (which tends to only work successfully for well-known places), or by estimating visual attributes that can help infer location (for example, that a photo depicts a beach scene, thus limiting the possible locations to coastlines). The former techniques tend to have very high precision, but low recall, whereas the latter techniques have much less precision (but higher recall).

One big challenge with all these approaches to extracting entities from images is dealing with the sheer amount of data required. For all the techniques, large amounts of image data are required to learn the visual representations. In some cases, this makes the problem intractable without additional constraints. For example, in the case of face recognition, or even logo recognition, it is not possible to have multiple images of all the people (or logos) in the world from which to train discriminative classifiers. Typically, these problems are constrained by deciding a priori specifically who (or what) needs to be detected in the images being analysed. Another way of constraining the analysis is to make use of any available information from the context of the image in question (for example analysis of surrounding text, titles, tags, etc.), and use this to guide the visual analysis.

4.2 Exploring Human Faces

Human faces are an obvious starting point for image analysis as they can potentially tell us who is in the image, as well as allowing us to make inferences about that person's emotional state. Before any higher-level analysis can occur, faces must first be detected in an image. The problem of face detection has been studied for a very long time in the computer vision field and, while not solved completely, has a number of acceptable solutions (under certain constraints, such as requiring the face to be "frontal" or approximately facing the camera).

While by no means the only (or best) approach, the algorithm for face detection developed by Viola and Jones [26] is probably the most widespread computer-vision technique of all time. Viola and Jones' technique works by training cascades of simple classifiers based on certain small patterns of light and dark pixels (these patterns are often referred to as Haar-like features, as they approximate the Haar wavelet function). When trained on large sets of human face images, the resultant classifier cascade can detect faces in images robustly and efficiently. In the case of human faces, the trained classifiers recognise patterns common to all faces, such as the areas directly above and below the eye generally having lighter intensity than the eye itself.

4.2.1 Analysing Facial Expression

Once a face has been localised, it is possible to make measurable estimates of that individual's facial expression in the image [27–29], as well as other attributes such as gender. Facial expressions are of particular interest because psychological studies have shown facial expressions can be used to infer the emotional state of the individual [30], and thus be used to infer sentiment. The Facial Action Coding System [31] (FACS) is a tool developed by psychologists to provide a standardised way of describing the expressions of faces. Codes represent muscular actions in the face (such as "inner eyebrow raising", or "lip corner puller"). Further coding systems such as EMFACS [32] and FACSAID [33] provide combinations of FACS codes that represent emotions (for example, activation of the lip corner puller AU6 and the cheek raiser AU12 actions imply happiness).

Given a detection of a human face in an image, it is possible to fit a flexible *shape* model that describes the overall intrinsic characteristics of the depicted individual's face and their expression, as well as extrinsic characteristics such as the pose of the person relative to the camera. Active Shape Models [34] (ASMs), Active Appearance Models [35] (AAMs) and Constrained Local Models [36] (CLMs) are well-studied algorithms for fitting a flexible shape to an image using the image's content to choose the best position for the vertices of the shape whilst constraining the shape to be plausible (based on a set of training examples that define the extents of the shape). As these models are both parametric (the shape is controlled by a small number of parameters) and generative (they allow a face to be reconstructed using these parameters), a large range of poses, expressions and appearances (skin textures) can be generated. Fitting a model to an image is a constrained optimisation problem in which the parameters of the model are iteratively updated in order to minimise the difference between the generated model and the image. Once a model is fitted to an image, the parameters can then be used as input to an expression classifier that can determine an expression label for the face. More specifically, the muscular movements encoded by FACS map to combinations of parameters in the face model, so a classifier can be potentially trained to recognise these actions [37–39]. Figure 2 shows a screenshot of our experimental CLM-based expression recognition system which has been trained to recognise FACS AUs in a laboratory setting, with highly constrained imaging conditions (i.e. restricted pose, uncluttered background, etc.).

Unfortunately, training a system to detect the full set of action units required for the different emotional states is difficult due to the lack of publicly available data. A second problem directly relates to the facial models themselves, in that it is quite difficult to build a shape model (ASM, AAM or CLM) that will accurately fit all faces, which is essential for the accurate measurement of the shape parameters needed for expression classification. A third and final problem is that accurate detection of a face is required to initialise the fitting of the model; whilst face detection techniques are quite mature, they can still have major problems working in real-world images where the faces are not exactly frontal to the camera, or there are shadows or contrast issues. Using real-world images collected from the web and social web, we found that inaccuracies in the face model alignment would regularly cause misclassification of

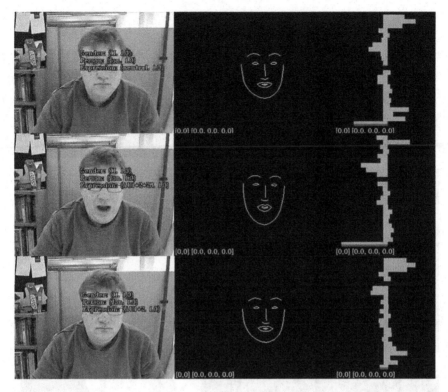

Fig. 2 Recognition of expressions in a laboratory setting using a CLM. The bars on the *right* illustrate the values of the parameter vector which define the shape of the model shown in the *centre*. Automated fitting techniques are used to adjust the values in the parameter vector so that the generated shape optimally matches the face in the image on the *left*

the action units, and therefore the expressions. Figure 3 shows some examples of the trained CLM model illustrated in Fig. 2 applied to example images collected from social media that are related to the US Elections. Notice in particular how poorly the model fits to Michelle Obama's face (and causes the misclassification of gender as a side effect). As this is a rapidly moving area of research, it will be interesting to see how expression modelling techniques develop over the coming years, especially in the presence of benchmarks such as Facial Expressions in the Wild[3] [40].

4.2.2 Recognising People

Once faces have been detected, recent advancements in face recognition mean that people can be recognised with relatively high accuracy from within a small search space (i.e., a relatively small set of people to choose from). The problem with a

---

[3]http://cs.anu.edu.au/few.

Fig. 3 Examples illustrating a CLM-based shape model with associated attribute classifiers applied to real images from the social web

general media analysis scenario is that the search space is effectively infinite, and current face recognition algorithms tend to deteriorate rapidly as the search space gets larger. One option that we have started to explore in our recent work is to apply entity recognition to any available contextual text to extract mentions of people, which we then use to constrain the face recogniser's search space to a small subset of person entities. For well-known personalities and people whose photos can be found on the

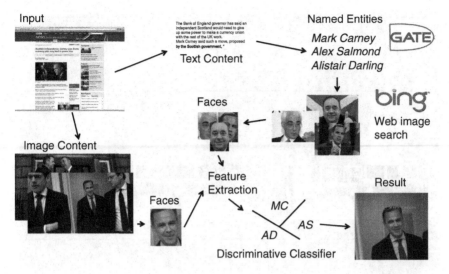

Fig. 4 Automated face verification using the names of people detected from the contextual text

Internet, a web-based image search can be used to automatically retrieve example images of those people from which a face recognition algorithm can be trained [41]. An illustration of the overall process used in our recent experiments is shown in Fig. 4. This overall process of using the contextual information to guide what to look for in the image is equally applicable to other types of entity, such as organisations with their corporate logos.

4.3 Contextualising Image Reuse

One way of gathering interesting insights into the social web is to look at how media spreads. In particular, we can measure how it is reused and talked about over time, and whether the aspects of the context, such as sentiment, change. One very powerful affordance gained from using near-duplicate images in this way is that the analysis is agnostic of the context, and in particular can be used to link together very different contexts which share the same image. From a practical point of view, duplicate images can be used to infer links between social media documents with text in a variety of human languages without the need to explicitly understand those languages.

Detecting duplicate images is not just a matter of looking at the url from which the media is hosted, because the same image is often hosted in many different locations, often with subtle (or not so subtle) changes from compression, cropping, rotation, etc. Using recent computer vision techniques, *near-duplicate* images can be detected efficiently across very large static datasets [42, 43] and streams of (social) media [44].

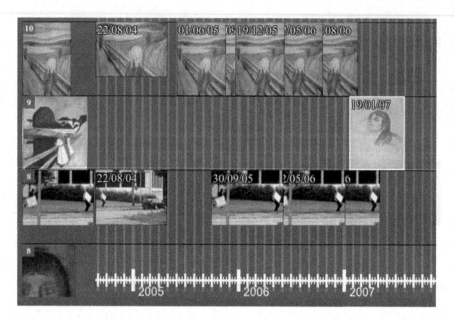

Fig. 5 Visualising how images are reused over time

The technology behind these systems varies, but typically relies on some form of robust image feature extraction followed by an indexing step to enable images to be efficiently compared. The SIFT local feature [45] is a popular choice to describe the image's content as it is highly robust to the typical transformations that make images near-duplicates rather than exact duplicates. For the indexing step, vector-quantisation followed by storage in an inverted index [43, 46], and locality sensitive hashing [42, 44, 47] are popular approaches.

4.3.1 Mining Temporal Reuse

Given a corpus of documents containing images in which we know the time that the document was created or posted, we can start to explore how a given image is reused over time. Figure 5 shows a screenshot of an experimental visualisation that displays duplicate images on a timeline, based on the date of the document that contained the image. From this visualisation it is possible to see how the incidence of the image varies over time as well as identifying clusters which may signify important time periods within the narrative of the image. In particular, in the specific case of the data used for the visualisation in Fig. 5 (which in this case was created from a web crawl) it is possible to see hidden patterns of reuse being exposed. The topmost band shows images of a painting called "The Scream" by Edvard Munch. In 2004 this painting was stolen from a museum in Norway and it is here where the image is first used. During the following 3 years, the story about the stolen painting appeared in news articles as the thieves were arrested and charged, and the painting then recovered; three separate events in the narrative of this story which are elucidated by the visualisation.

Interestingly, the example shown in Fig. 5 also displays a time correlation between the picture of The Scream and the picture two lines below. This second picture is a photograph of the thieves making off with the painting itself. This correlation can be investigated by looking at the contextual information from the document in which the image was embedded; in this case, the correlation is, perhaps, expected as the photograph is related to the story of the stolen painting. However, the stories to which that photograph is related are very different to those to which the picture of the painting are related, despite the correlation. Indeed, examining the narrative thread exposed by the visualisation makes it clear that the picture of the painting is associated with the narrative of the painting being stolen, whereas the photograph of the thieves is associated with complementary articles about protecting museum artefacts.

4.3.2 Mining Sentiment and Opinion Polarities from Reused Images

Recently, we developed a system called Twitter's Visual Pulse [44] which finds near-duplicate images over fixed time periods in a live Twitter stream. By extracting the sentiment from the tweets associated with these duplicate images (using the techniques in Sect. 3), we can find out how the image is used in different contexts. In many cases, the image may be reused in contexts which are, overall, sentimentally ambivalent; however, there may be cases where an image is used in a consistent way—for example, a particular image may be used in consistently positive tweets. We form a discrete probability distribution for images falling in specific sentiment categories, which we can use to assign sentiment probabilities to the image when it is further reused, particularly in cases where the textual sentiment analysis is inconclusive. When a context has conflicting opinions, or an opinion is not evident, then the image may be able to provide clues as to the article's sentiment: if it contains an image which has been reused many times in articles that have particular opinions, the ambiguous article can be associated with that opinion through the association with the image. Because the image matching is purely visual, this technique will work across language barriers, such that articles in a language that cannot be analysed could still have sentiment scores associated with them.

It can also be instructive to visualise the sentiments or opinion polarities associated with the contexts of particular images as they are reused. Figure 6 shows an example of this.

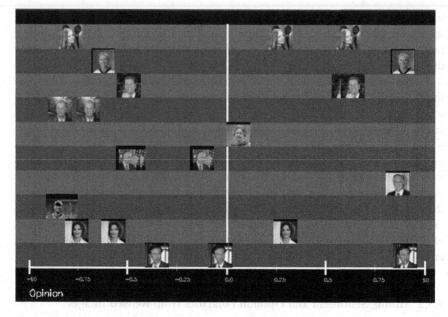

Fig. 6 Visualising how reused images vary with respect to the opinion polarity of their context

4.4 Exploring Multimodal Sentiment, Privacy and Attractiveness in Social Images

Opinion and sentiment are rather complex notions that can be very difficult to predict purely from visual data alone. A more fruitful approach is to consider the image (or other media modality) in the context in which it appears, whether that be an image on Flickr or video on YouTube surrounded by tags and comments provided by humans; or an image in a news item surrounded by the text of the article to which it relates. State-of-the-art research on the sentiment analysis of images (see e.g. [15, 16, 48–50]) has already begun to explore how the analysis of textual content and the analysis of visual content can complement each other. Recently, we have been exploring how visual content and contextual information can be leveraged to train machines to predict facets related to opinion formation.

4.4.1 Image Sentiment

In less constrained multimedia, we cannot rely on there being faces in the images, and sentiment may be carried by other visual traits. Indeed, images may intrinsically have sentiment associated with them through design (such as a poster for a horror film) or through association with a specific subject matter which may be context sensitive (such as a photo of wind generators in the context of climate change). For these situations there are no specific algorithms we can use for extracting the sentiment. However, we can look for correlations between visual features and textual labels using classifiers and regressors trained over ground-truth datasets. Unfortunately, large, well labelled datasets for image sentiment are hard to come by. For that reason, we turned to user-provided image annotations to generate a large dataset to use for classification. Using SentiWordNet [51], we queried Flickr for the words that had the strongest positive and negative sentiments, and retrieved sets of images for each of them. Combined, these formed a ground-truth for positive and negative sentiment images. Full details of the dataset and the trained classifiers are described in [49], but we will summarise the conclusions here.

We gathered images for the 1000 strongest sentiment words from SentiWordNet. This resulted in 586,000 images, most of which had a resolution of more than 1 megapixel. We extracted global and local colour features (these describe the colour distribution in the image, and in the case of the local variant, a coarse spatial layout of the colour distribution) and SIFT local features [45] (which describe small patches of texture/pattern in the image) from the images. Using these features a linear SVM classifier was trained to recognise positive/negative sentiment. We observed that for small recall values, precision values of up to 70 % can be reached. Due to the challenging character of this task, for high recall values, the precision degrades down to the random baseline. Interestingly, using mutual information, we were able to reverse engineer the correlations in the classifier to determine which features were correlated to which labels. We found that positive images had overall warm colours (reds, oranges, yellows, skin tones) and negative images had colder colours (blues, dark greens). The location of the colour had no real significance. The negative SIFT features seem dominated by a very light central blob surrounded by a much darker background, while the positive SIFT features are dominated by a dark blob on the side of the patch.

4.4.2 Image Privacy

In terms of privacy classification, we have been able to construct classifiers using textual tags and visual features, both combined and separately, in order to predict whether an image is potentially of a private nature. This is directly related to opinion formation, because it can potentially be used to identify images such as paparazzi shots and leaked private images which have been published or posted in public places. For our privacy classification experiments [52, 53], we created a dataset of 90,000 "recently uploaded" images from Flickr with a minimum of 5 English tags. In order to create ground-truth, we created a social annotation game and used crowdsourcing to get the opinions of multiple individuals. In the game, users were able to select three different options for each image they were presented with: private, undecidable or public. Users were given the following advice before commencing the game: *Private photos are photos which have to do with the private sphere (like self portraits, family, friends, your home) or contain objects that you would not share with the entire world (like a private email). The rest is public. In case no decision can be made, the picture should be marked as undecidable.*

Altogether the participants annotated 83,820 images. Analysis showed that around 78 % of photos were labeled as public or undecidable by all of the participating judges. This is to be expected due to the nature of images on Flickr, which are on the whole posted to be shared with the public at large. From the remaining 22 % of photos, 12 % were labeled as "private" by all the judges, and 10 % received "private" votes from at least one of the judges. A subsample of the data with the highest annotator agreement was selected for performing classification experiments.

A selection of different visual features were extracted from the images for training input to linear SVM classifiers. The textual feature was a simple word-occurrence histogram, with stemming applied to the tags to reduce variability and group similar tags. The classifiers were created and evaluated for each individual feature, all visual features combined, and text and visual features combined. Combined features worked better than individual features; evaluation using precision-recall metrics showed a break-even point of 0.74 for visual features, 0.78 for textual features and 0.80 for combined text and visual features.

4.4.3 Image Attractiveness

When considered within the context of the article or post in which it appears, we hypothesise that the attractiveness of a photograph can be a strong indicator of the opinion and sentiment expressed by the article. Currently, we are only beginning to scratch the surface of this area, but we have been investigating building computational models of attractiveness that take into account both visual features as well as surrounding contextual tags [54].

On the assumption that on Flickr, more attractive or aesthetically pleasing photographs have higher numbers of favourite assignments, we built a dataset of 400,000 images as follows. We randomly selected time periods of 20 min from a time span of 5 years 2005–2010. From each of the periods we selected at most 5 pictures from Flickr with the highest number of favourite assignments as positive examples, as well as the same number of photos without favourite assignments as negative examples. We stopped after obtaining a set of 200,000 photos from each class.

Even though aesthetic and artistic quality cannot be quantitatively computed, it has been shown that certain visual features of images have significant correlation with them. For instance, appealing images tend to have higher colourfulness, increased contrast and sharpness [55]; we apply image analysis to extract these features. Bag-of-words textual features extracted from the title and tags can also provide information about the image quality and aesthetics. By training linear SVM classifiers, we are able to generate predictive models of image attractiveness using these features. Experiments (see [54] for full details) have shown that our visual features can provide reasonable performance (break-even-point of 0.67 with respect to the precision-recall curve), whilst combinations of the textual and visual features perform better than either the textual or visual features alone (combined feature break-even-point of 0.84).

5 Conclusions

In this paper, we have described the general approach we undertake to the analysis of social media, using a combination of textual and multimedia opinion mining tools. It is clear that both opinion mining in general, and the wider analysis of social media, are difficult tasks from both perspectives, and there are many unresolved issues. The modular nature of our approach also lends itself to new advances in a range of subtasks: from the difficulties of analysing the noisy forms of language inherent in tweets, to the problems of dealing with sarcasm in social media, to the ambiguities inherent in such forms of web content that inhibit both textual and multimedia analysis tools. Furthermore, to our knowledge this is the first system that attempts to combine such kinds of textual and multimedia analysis tools in an integrated system, and preliminary results are very promising, though this is nevertheless very much ongoing research. Future work includes further development of the opinion mining tools: we have already begun investigations into issues such as sarcasm detection, more intricate use of discourse analysis and so on.

Acknowledgments This work was supported by the European Union under grant agreements No. 270239 Arcomem (http://www.arcomem.eu) and No. 610829 DecarboNet (http://www.decarbonet.eu).

References

1. Taboada, M., Brooke, J., Tofiloski, M., Voll, K., Stede, M.: Lexicon-based methods for sentiment analysis. Comput. Linguist. **1**, 1–41 (2011)
2. Boiy, E., Moens, M.F.: A machine learning approach to sentiment analysis in multilingual web texts. Inf. Retr. **12**, 526–558 (2009)
3. Moghaddam, S., Popowich, F.: Opinion polarity identification through adjectives. CoRR arXiv:1011.4623 (2010)
4. Mullaly, A., Gagné, C., Spalding, T., Marchak, K.: Examining ambiguous adjectives in adjective-noun phrases: evidence for representation as a shared core-meaning. Mental Lexicon **5**, 87–114 (2010)
5. Weichselbraun, A., Gindl, S., Scharl, A.: A context-dependent supervised learning approach to sentiment detection in large textual databases. J. Inf. Data Manage. **1**, 329–342 (2010)
6. Gindl, S., Weichselbraun, A., Scharl, A.: Cross-domain contextualisation of sentiment lexicons. In: Proceedings of 19th European Conference on Artificial Intelligence (ECAI-2010), pp. 771–776 (2010)
7. Pak, A., Paroubek, P.: Twitter based system: using Twitter for disambiguating sentiment ambiguous adjectives. In: Proceedings of the 5th International Workshop on Semantic Evaluation, pp. 436–439 (2010)

8. Go, A., Bhayani, R., Huang, L.: Twitter sentiment classification using distant supervision. Technical Report CS224N Project Report, Stanford University (2009)
9. Derczynski, L., Maynard, D., Aswani, N., Bontcheva, K.: Microblog-Genre noise and impact on semantic annotation accuracy. In: Proceedings of the 24th ACM Conference on Hypertext and Social Media, ACM (2013)
10. Liu, X., Zhang, S., Wei, F., Zhou, M.: Recognizing named entities in tweets. In: Proceedings of the 49th Annual Meeting of the Association for Computational Linguistics: Human Language Technologies, pp. 359–367 (2011)
11. Tsur, O., Davidov, D., Rappoport, A.: Icwsm-a great catchy name: semi-supervised recognition of sarcastic sentences in online product reviews. In: Proceedings of the Fourth International AAAI Conference on Weblogs and Social Media, pp. 162–169 (2010)
12. Liebrecht, C., Kunneman, F., van den Bosch, A.: The perfect solution for detecting sarcasm in tweets# not. WASSA **2013**, 29 (2013)
13. Reyes, A., Rosso, P., Veale, T.: A multidimensional approach for detecting irony in Twitter. In: Language Resources and Evaluation, pp. 1–30 (2013)
14. Davidov, D., Tsur, O., Rappoport, A.: Semi-supervised recognition of sarcastic sentences in Twitter and Amazon. In: Proceedings of the Fourteenth Conference on Computational Natural Language Learning, Association for Computational Linguistics, pp. 107–116 (2010)
15. Yanulevskaya, V., Van Gemert, J., Roth, K., Herbold, A.K., Sebe, N., Geusebroek, J.M.: Emotional valence categorization using holistic image features. In: 15th IEEE International Conference on Image Processing, 2008. ICIP 2008, pp. 101–104 (2008)
16. Wei-ning, W., Ying-lin, Y., Sheng-ming, J.: Image retrieval by emotional semantics: a study of emotional space and feature extraction. In: IEEE International Conference on Systems, Man and Cybernetics, 2006. SMC '06, vol. 4, pp. 3534–3539 (2006)
17. Hare, J.S., Lewis, P.H., Enser, P.G.B., Sandom, C.J.: A linear-algebraic technique with an application in semantic image retrieval. In Sundaram, H., Naphade, M.R., Smith, J.R., Rui, Y. (eds.) CIVR. Lecture Notes in Computer Science, vol. 4071, pp. 31–40. Springer, New York (2006)
18. Bontcheva, K., Derczynski, L., Funk, A., Greenwood, M.A., Maynard, D., Aswani, N.: TwitIE: an open-source information extraction pipeline for microblog text. In: Proceedings of the International Conference on Recent Advances in Natural Language Processing, Association for Computational Linguistics (2013)
19. Maynard, D., Bontcheva, K., Rout, D.: Challenges in developing opinion mining tools for social media. In: Proceedings of @NLP can u tag #usergeneratedcontent?! Workshop at LREC 2012, Turkey (2012)
20. Miller, G.A., Beckwith, R., Felbaum, C., Gross, D., Miller, C.Miller, G.A., Beckwith, R., Felbaum, C., Gross, D., Miller, C., Minsky, M.: Five papers on WordNet (1990)
21. Maynard, D., Greenwood, M.A.: Who cares about sarcastic tweets? Investigating the impact of sarcasm on sentiment analysis. In: Proceedings of LREC 2014, Reykjavik, Iceland (2014)
22. Berardi, G., Esuli, A., Marcheggiani, D., Sebastiani, F.: ISTI@ TREC microblog track 2011: exploring the use of hashtag segmentation and text quality ranking. In: TREC (2011)
23. Maynard, D., Gossen, G., Fisichella, M., Funk, A.: Should I care about your opinion? Detection of opinion interestingness and dynamics in social media. J. Future Internet (in press)
24. Kalantidis, Y., Pueyo, L.G., Trevisiol, M., van Zwol, R., Avrithis, Y.: Scalable triangulation-based logo recognition. In: Proceedings of the 1st ACM International Conference on Multimedia Retrieval. ICMR '11, pp. 20:1–20:7, New York, NY, USA, ACM (2011)
25. Psyllos, A., Anagnostopoulos, C.N., Kayafas, E.: M-sift: a new method for vehicle logo recognition. In: 2012 IEEE International Conference on Vehicular Electronics and Safety (ICVES), pp. 261–266 (2012)
26. Viola, P., Jones, M.: Robust real-time object detection. In: International Journal of Computer Vision (2001)
27. Fasel, B., Luettin, J.: Automatic facial expression analysis: a survey. Pattern Recogn. **36**, 259–275 (2003)

28. Pantic, M., Sebe, N., Cohn, J.F., Huang, T.: Affective multimodal human-computer interaction. In: Proceedings of the 13th annual ACM international conference on Multimedia. MULTIME-DIA '05, pp. 669–676, New York, NY, USA, ACM (2005)
29. Tian, Y.l., Kanade, T., Cohn, J.F.: Facial expression analysis. Handbook of Face Recognition, pp. 247–275 (2005)
30. Du, S., Tao, Y., Martinez, A.M.: Compound facial expressions of emotion. Proc Natl Acad Sci USA (2014)
31. Ekman, P., Friesen, W.: Facial Action Coding System: A Technique for the Measurement of Facial Movement. Consulting Psychologists Press, Palo Alto (1978)
32. Friesen, W., Ekman, P.: EMFACS-7: Emotional Facial Action Coding System. University of California, California (1983) (Unpublished manual)
33. Ekman, P., Irwin, W., Rosenberg, E.R., Hager, J.C.: FACS Affect Interpretation Database (FACSAID). http://face-and-emotion.com/dataface/facsaid/description.jsp (1997)
34. Cootes, T.F., Taylor, C.J., Cooper, D.H., Graham, J.: Active shape models—their training and application. Comput. Vis. Image Underst. **61**, 38–59 (1995)
35. Cootes, T., Edwards, G., Taylor, C.: Active appearance models. IEEE Trans. Pattern Anal. Mach. Intell. **23**, 681–685 (2001)
36. Saragih, J.M., Lucey, S., Cohn, J.: Face alignment through subspace constrained mean-shifts. In: International Conference of Computer Vision (ICCV) (2009)
37. Lucey, P., Cohn, J., Kanade, T., Saragih, J., Ambadar, Z., Matthews, I.: The extended cohn-kanade dataset (ck+): a complete dataset for action unit and emotion-specified expression. In: 2010 IEEE Computer Society Conference on Computer Vision and Pattern Recognition Workshops (CVPRW), pp. 94–101 (2010)
38. Chew, S., Lucey, P., Lucey, S., Saragih, J., Cohn, J., Sridharan, S.: Person-independent facial expression detection using constrained local models. In: 2011 IEEE International Conference on Automatic Face Gesture Recognition and Workshops (FG 2011), pp. 915–920 (2011)
39. Ryan, A., Cohn, J.F., Lucey, S., Saragih, J., Lucey, P., De la Torre, F., Ross, A.: Automated facial expression recognition system. In: IEEE International Carnahan Conference on Security Technology (2009)
40. Dhall, A., Goecke, R., Lucey, S., Gedeon, T.: Static facial expression analysis in tough conditions: data, evaluation protocol and benchmark. In: 2011 IEEE International Conference on Computer Vision Workshops (ICCV Workshops), pp. 2106–2112 (2011)
41. Parkhi, O., Vedaldi, A., Zisserman, A.: On-the-fly specific person retrieval. In: 13th International Workshop on Image Analysis for Multimedia Interactive Services (WIAMIS), pp. 1–4 (2012)
42. Dong, W., Wang, Z., Charikar, M., Li, K.: High-confidence near-duplicate image detection. In: ACM ICMR'12, pp. 1:1–1:8, ACM (2012)
43. Hare, J., Samangooei, S., Dupplaw, D., Lewis, P.: Imageterrier: an extensible platform for scalable high-performance image retrieval. In: ICMR 2012 (2012)
44. Hare, J., Samangooei, S., Dupplaw, D., Lewis, P.H.: Twitter's visual pulse. In: 3rd ACM International Conference on Multimedia Retrieval, pp. 297–298 (2013)
45. Lowe, D.: Distinctive image features from scale-invariant keypoints. IJCV **60**, 91–110 (2004)
46. Sivic, J., Zisserman, A.: Video google: a text retrieval approach to object matching in videos. In: ICCV, pp. 1470–1477 (2003)
47. Dong, W., Charikar, M., Li, K.: Asymmetric distance estimation with sketches for similarity search in high-dimensional spaces. In: SIGIR'08, ACM, pp. 123–130 (2008)
48. Zontone, P., Boato, G., Hare, J., Lewis, P., Siersdorfer, S., Minack, E.: Image and collateral text in support of auto-annotation and sentiment analysis. In: TextGraphs-5: Graph-based Methods for Natural Language Processing, The Association for Computational Linguistics, pp. 88–92 (2010)
49. Siersdorfer, S., Hare, J., Minack, E., Deng, F.: Analyzing and predicting sentiment of images on the social web. In: ACM Multimedia 2010, pp. 715–718, ACM (2010)
50. Wang, W., He, Q.: A survey on emotional semantic image retrieval. In: 15th IEEE International Conference on Image Processing, 2008. ICIP 2008, pp. 117–120 (2008)

51. Esuli, A., Sebastiani, F.: SENTIWORDNET: a publicly available lexical resource for opinion mining. In: Proceedings of LREC 2006 (2006)
52. Zerr, S., Siersdorfer, S., Hare, J., Demidova, E.: Privacy-aware image classification and search. In: SIGIR'12, pp. 35–44, ACM, New York, NY, USA (2012)
53. Zerr, S., Siersdorfer, S., Hare, J.: Picalert!: a system for privacy-aware image classification and retrieval. In: 21st ACM Conference on Information and Knowledge Management (CIKM 2012) (2012)
54. Siersdorfer, S., Zerr, S., Pedro, J.S., Hare, J.: Nicepic!: a system for extracting attractive photos from flickr streams. In: ACM SIGIR 2014, ACM (2014)
55. Pedro, J.S., Siersdorfer, S.: Ranking and classifying attractiveness of photos in folksonomies. In: 18th International World Wide Web Conference, pp. 771–771 (2009)

Context-Aware Sentiment Analysis of Social Media

Aminu Muhammad, Nirmalie Wiratunga and Robert Lothian

Abstract The lexicon-based approach to opinion mining is typically preferred where training data is difficult to obtain or cross domain robustness of algorithms is of essence. However, this approach suffers from the semantic gap between the polarity with which a sentiment-bearing term appears in the text (i.e. contextual polarity) and its prior polarity captured by the lexicon. This is further exacerbated when mining is applied to social media. Here, we propose an approach to address this semantic gap. Firstly, by accounting for the influence of surrounding terms to a sentiment bearing term (local context). Secondly, by accounting for content and context disagreement between the lexicon and the domain in which it is applied (global context). This is achieved by generating a domain-focused lexicon using distant-supervision and integrating its scores with a generic lexicon (SentiWordNet). Evaluation results from sentiment classification over social media content extracted from three different platforms show benefits of accounting for local and global contexts, both individually and in combination. We also present some promising results from our investigation into the cross-platform transferability of our approach.

1 Introduction

Sentiment analysis or opinion mining concerns the study of opinions expressed in text. Typically, an opinion comprises of its polarity (positive or negative), the target (or specific aspects of the target) to which the opinion was expressed and the time at which the opinion was expressed [14]. Applications of Sentiment analysis have been established in the areas of politics [3], stock markets [1], economic systems [15] and

A. Muhammad (✉) · N. Wiratunga · R. Lothian
IDEAS Research Institute, Robert Gordon University, Aberdeen, Scotland, UK
e-mail: a.b.muhammad1@rgu.ac.uk

N. Wiratunga
e-mail: n.wiratunga@rgu.ac.uk

R. Lothian
e-mail: r.m.lothian@rgu.ac.uk

© Springer International Publishing Switzerland 2015
M.M. Gaber et al. (eds.), *Advances in Social Media Analysis*,
Studies in Computational Intelligence 602,
DOI 10.1007/978-3-319-18458-6_5

security concerns [11] among others. For instance, a business would want to know customer's opinion about its products/services and that of its competitors. Similarly, decision making can be shaped by stakeholder opinions (reviews or comments) [14].

Aggregation of sentiment polarity scores from a resource such as sentiment lexicon is typically used to classify opinionated text into sentiment classes. As a results several general purpose sentiment lexicons have been developed and made public for research e.g. General Inquirer [26], Opinion Lexicon [10] and SentiWordNet [2]. However, performance of lexicon-based sentiment analysis still remains below acceptable levels. This is because the polarity with which a sentiment-bearing term appears in text (i.e. contextual polarity) can be different from its prior polarity offered by a lexicon. Two forms of semantic difference seems to contribute to this semantic gap. First, difference in *local context* arising from the interaction of a sentiment-bearing term with its textual surrounding. For example, the prior polarity of 'good' is positive, however, such polarity is changed in 'not good'. Second, the difference in *global context* arising from the difference in the typical sense of a term captured by a lexicon and the term's domain-specific usage. For example, in the text 'the movie sucks', although the term 'sucks' seems highly sentiment-bearing, this may not be reflected by a general purpose sentiment lexicon. Another problem with typical sentiment lexicons is that they cannot adapt to changes in vocabulary usage which is particularly evident in social media.

In this work, we propose an approach to accounting for local and global contexts in social media domains. First, we introduce strategies to account for sentiment modifiers: negations, intensifiers and diminishers. This also includes non-lexical modifiers commonly used to express or emphasise sentiment in social media: capitalisation, sequence of repeated character and emoticons. Second, we address the problem content (vocabulary) and context between a sentiment lexicon and its domain of application. Here, we use distant-supervision [9, 24] to mine sentiment knowledge from the target domain and thereafter combine this with knowledge obtained from SentiWordNet. The remainder of this paper is organised as follows. Section 2 describes related work. Sentiment analysis using SentiWordNet is discussed in Sect. 3, followed by our approach to account for local and global contexts in Sects. 4 and 5. Evaluation and discussions appear in Sect. 6, followed by conclusions and future work in Sect. 7.

2 Related Work

Broadly, three methods have been employed for sentiment classification namely machine learning, lexicon based and hybrid. For machine learning, supervised classifiers are trained with sentiment labelled data commonly generated through labour-intensive human annotation. The trained classifiers are then used to classify new documents for sentiment. Machine learning classifiers such as Naïve Bayes (NB), Maximum Entropy (ME) and Support Vector Machines (SVMs) have been used for sentiment classification [21]. Here, results show that, like topic-based text classification, SVMs perform better than NB and ME. However, performance of all classifiers on sentiment is lower than in topic-based classification alluding to the fact that sentiment classification is more challenging than topic classification. Unlike with topic, sophisticated representation schemes that go beyond bag-of-words is needed: for instance like term frequencies [20]; appraisal groups [30] and feature subsumption hierarchies [25].

Access to labelled data is of particular concern for sentiment classification when applied to social media (e.g. discussion forums, blogs and tweets). Recently distant-supervision was proposed to address this problem [9, 24]. Here, emoticons supplied by authors of tweets were used as noisy sentiment labels. Evaluation results on NB, ME and SVMs trained with distant-supervised data but tested on hand-labelled data show the approach to be effective with ME attaining the highest accuracy of 83 % on a combination of unigram and bigram features. Although distant-supervision has addressed the labelling problem for machine learning; the need for frequent re-learning and poor transferability across domains arguably makes lexicon based approaches better suited for social media content.

The lexicon based method excludes the need for labelled training data but requires a sentiment lexicon (i.e. a dictionary that associate terms with sentiment scores). Lexicon-based sentiment classification begins with the creation of a sentiment lexicon or the adoption of an existing one, from which sentiment scores of terms are extracted and aggregated to predict sentiment of a given piece of text. Such lexicons are either manually generated or semi-automatically generated from generic knowledge sources. Manually generated lexicons are obviously more accurate, however, they tend to have relatively smaller term coverage. For example in General Inquirer [26] and Opinion Lexicon [10] whilst sentiment polarity scores were assigned by humans, they cover only 4216 and 6789 terms respectively. In contrast semi-automatically generated lexicons such as the corpus-based in [16] and dictionary-based SentiWordNet [2] each covers over 20,000 words. However, poor coverage remains a problem when applied to social media.

Early work in lexicon-based sentiment analysis are based on the aggregation of individual polarities of terms irrespective of grammatical dependencies that may exist between them. This approach is incomplete and often gives the wrong results when implemented directly because term prior polarity changes due to the effect of other terms with which the term co-occurs. Contextual analysis due to terms co-occurrence with contextual valence shifters was subsequently introduced [12, 23]. Here, polarities of sentiment-bearing terms that are under the influence of *negation* terms (e.g. *'not'*, *'never'*, *'nothing'*) is inverted and that of those terms that are under the influence of terms that increase the polarity strength of sentiment terms (i.e. intensifiers e.g. *'very'*, *'highly'*) and terms that decrease the polarity (i.e. diminishers e.g. *'slightly'* and *'a-little-bit'*) are increased and decreased respectively. Sentiment negation analysis is of particular challenge as the polarity of negated term does not always translate to its opposite. For instance, whereas "It is *not good*" is more or less the same as "It is *bad*", "It is *not excellent*" is more positive than "It is *horrible*". Consequently, *shift* approach was proposed as a preferred alternative to sentiment inversion for negation [27]. Here, prior polarity of sentiment terms that are under the influence of negation terms is inverted, and also reduced by a certain weight. We use the shift approach for negation in this work and introduce strategies to address the effect of intensifiers and diminishers but unlike the previous work, we use a lexicon in which terms are associated to both positive and negative polarities.

Social media presents additional challenge to sentiment analysis as users often use non-standard but creative means to express sentiment. In more recent work, lexicon-based sentiment analysis was extended to incorporate modification of term prior polarities based non-lexical modifiers [19, 28, 29]. Such non-lexical modifiers include term elongation by repeating character (e.g. 'haaappppyy' in place 'happy'), capitalization of terms, and internet slang. We also implement strategy to account for non-lexical modifiers. However, semantic gap between prior and contextual polarities still remains in social media due its evolving vocabulary and context of terms.

The hybrid method to sentiment analysis involves combining advantageous aspects from lexicon-based and machine learning in a complementary manner. For instance, sentiment scores of terms obtained from a lexicon are used as additional features to train machine learning classifiers [5, 18]. Also, machine learning is employed to adapt the lexicon to social media content [29], where initial scores for terms, assigned manually are increased or decreased based on observed classification accuracies on a manually labelled dataset. In our work we also advocate the hybrid approach, however we employ distant-supervision as a means to enhance an existing lexicon as opposed to combining with a classifier or manually labelled data. In this way, we improve on coverage and context, avoiding the need for re-learning yet maintaining transferability across domains.

3 Sentiment Analysis Using SentiWordNet

SentiWordNet is a general purpose lexicon for sentiment analysis tasks developed from WordNet [8]. Each WordNet's synset (i.e. a group of synonymous terms on a particular meaning) is associated with positive and negative scores indicating the synset's association with positive and negative sentiment classes respectively. An objective score can also be inferred as the difference of subtracting positive and negative scores from 1. This score is especially useful for determining whether or not a piece of text conveys any sentiment (i.e. subjectivity analysis). To generate this lexicon, a small manually labelled seed synsets were iteratively expanded by adding related synsets through *synonymy* and *antonymy* relations; whereby synonymy preserves while antonymy reverses polarity with a given seed synset. As there is no direct synonym relation between synsets in WordNet, the relations: see_also, similar_to, pertains_to, derived_from and attribute are used to represent the synonymy relation while direct antonym relation is used for antonymy. Thereafter, textual definitions and/or example sentences (i.e. glosses) from these synsets along with that of another set assumed to be composed of objective synsets were used to train eight ternary classifiers. The classifiers then classify every synset and the proportion of classification for each class (positive, negative and objective) were deemed to be the scores for the synset. In an enhanced version of the lexicon [7], the scores are optimised by

PoS	ID	+score	-score	synset	gloss
a	00016756	0	0.25	scarce#1	deficient in quantity or number compared with the demand; ...
n	00735936	0	0.625	misdeed#1 misbehaviour#1 misbehavior#1	improper or wicked or immoral behavior
r	00309249	0.125	0.125	despicably#1	in a despicable manner; "he acted despicably"
a	00017782	0.625	0	acceptable#1	worthy of acceptance or satisfactory; "acceptable levels of radiation" ...
n	04632063	0.75	0	chirpiness#1	cheerful and lively
v	02746140	0.625	0	beat#12	be superior; "Reading beats watching television"; "This sure beats work!"

Fig. 1 A fragment from SentiWordNet

the PageRank [4] algorithm. This starts with a manually selected synsets and then propagates sentiment polarity (positive or negative) is propagated to a target synset by assessing the synsets that connect to the target synset through the appearance of their terms in the gloss of the target synset.

Figure 1 shows a fragment from SentiWordNet. Scores for a synset or a specific word sense (word#sense) can be extracted by specifying the synset's ID or the word's lemma, part-of-speech (PoS) and sense number. Although scores are associated with a word sense, disambiguation is usually not performed as it does not seem to yield better results than using either the average score across all senses of a term at PoS level or the score attached to the most frequent sense of the term [6, 18, 22]. The baseline algorithm in this paper uses SentiWordNet (Algorithm 1). Given a document to be classified, Doc, it is first tokenized and lemmatized to generate a set of tokens, t_i. Each t_i is also assigned a PoS. Here, we use Stanford CoreNLP library.[1] Next, positive (t_i^+) and negative (t_i^-) scores for each term are extracted from the lexicon (step 3). Thereafter, these scores are respectively summed for all terms contained in Doc (steps 4–7 in Algorithm 1). Sentiment class of the input text is deemed positive if the normalised net positive score (Doc$^+$) exceeds the normalised net negative score (Doc$^-$) and negative otherwise (steps 8–13).

4 Local Context

We extend the BASE algorithm to account local context by considering neighbouring terms and inflection of terms for emphasis. In this extension (see Algorithm 2), prior polarity of a sentiment-bearing term in a document is modified depending on the term's proximity with lexical or non-lexical valence shifters.

---

[1]http://nlp.stanford.edu/software/corenlp.shtml.

Algorithm 1 BASE

INPUT: Doc, document to be classified
              S,   Sentiment Lexicon
OUTPUT: Class, Sentiment class for Doc
1: Initialise: Doc^+, Doc^-
2: **for all** $t_i \in$ Doc **do**
3:    Retrieve t_i^+ and t_i^- from S
4:    **if** $t_i^+ + t_i^- > 0$ **then**
5:        $Doc^+ \leftarrow Doc^+ + t_i^+$; $Doc^- \leftarrow Doc^- + t_i^-$
6:    **end if**
7: **end for**
8: Normalize Doc^+ and Doc^-
9: **if** $Doc^+ \geq Doc^-$ **then**
10:    **Return** *Positive*
11: **else**
12:    **Return** *Negative*
13: **end if**

4.1 Lexical Valence Shifters

Lexical valence shifters are sentiment modifiers that are in the form of known dictionary words. These are typically used to increase sentiment (i.e. intensifiers e.g. 'very', 'highly'); decrease sentiment (i.e. diminishers e.g. 'slightly', 'somewhat') or negate sentiment (i.e. negation terms, e.g. 'not', 'never'). Lexical valence shifters are associated with sentiment scores in SentiWordNet. For example, positive and negative scores of the adverb 'very' are 0.25 and 0.0 respectively, thus, the term always contribute score to positive dimension only. However, this term can also be used to increase negative sentiment, for example in 'very bad'. It is therefore imperative to identify the polarity affected by such modifiers and accordingly modify their scores. To achieve this, when an intensifier or deminisher is detected within the neighbourhood of a sentiment-bearing term, it is decoupled from scores offered by SentiWordNet and the dominant prior polarity (i.e. maximum of positive and negative scores) of the target sentiment-bearing term is increased (in the case of intensifier) or decreased (in the case of diminisher) relative to the strength of the intensifier or diminisher (steps 13–14, in Algorithm 2). We use a lexicon of intensifiers and diminshers [28] where each term is assigned strength of 1 or 2 indicating the degree to which the term increases or decreases sentiment. For instance, the intensification strength of 'extremely' is 2 while that of 'very' is 1. We convert these strengths to percentage increase or decrease in dorminant polarity of terms (50 % for 1 and 100 % for 2).

Figure 2 shows an example text containing an intensifier classified using BASE and our approach. The text, which is clearly negative is incorrectly classified as positive using sentiment scores of the intensifier (i.e. BASE approach). However, it is correctly classified using the intensification strength of the intensifier (i.e. our approach). Negation terms, however, are not only modifiers sentiment but also sentiment-bearing [17].

Text:	it's	really	awful		
Positive Score:	0.0	0.438	0.25		
Negative Score:	0.0	0.065	0.542		
BASE:					
Positive Score:	0.0 +	0.438 +	0.25 ⎤	Class =	
Negative Score:	0.0 +	0.065 +	0.542 ⎦	Positive	
EXTENDED:					
Positive Score:	0.0 +	0.0 +	0.25	⎤ Class =	
Negative Score:	0.0 +	0.0 +	0.542 x (100% + 50%)	⎦ Negative	

Fig. 2 Local context: Lexical valence shifter example

Accordingly we treat negation terms as both modifiers and sentiment-bearing. We use shift strategy to account for the modifier effect of negation [27] (step 12, in Algorithm 2). Here, when a negation term is detected within the neighbourhood of a sentiment-bearing term, positive and negative scores of the sentiment-bearing term are interchanged and reduced by a weight of $\frac{1}{5}$. The reduction in strength is to account for the fact that sometimes negation only reduces sentiment rather than complete inversion. Unlike with intensifiers and diminishers, lexicon scores associated with negation terms are retained.

4.2 Non-Lexical Valence Shifters

In addition to lexical valence shifters, non-lexical modifiers are also commonly used to increase sentiment. We introduce sentiment scores modification based on term inflection with a sequence of repeating characters/letters and capitalization. Here, the inflected term is identified and its dominant polarity (i.e. the maximum of positive and negative scores) is increased by the weight of 'very' (steps 4–7 and 17–18, in Algorithm 2). We chose the term 'very' as it is a typical intensifier. For capitalisation, the inflected term is simply lower-cased to get the original term. For a sequence of repeating letters or capitalization, the original term is identified by first reducing the number of the letter to a maximum of two and check with dictionary. If the intermediate word is not found, the repeating letters are further reduced to one letter, one sequence at a time. The occurrence of three or more consecutive exclamation or question marks or a mixture of both is treated as sentiment intensification of dominant polarities of the terms that are in the neighbourhood (a 5-token window before and after the multiple exclamation/question marks). Table 1 shows example of inflected terms using non-lexical valence shifters and how we resolve such inflection in this work. Notice that in the first row only the positive score of 'good' is increased

Table 1 Non-lexical valence shifters

Inflection	Equivalent	Aggregation
gooooood	Very good	$good^+ + [1 + Strength(very)]$
... BAD ...	Very bad	$bad^- + [1 + Strength(very)]$
Happy!!!	Very happy	$happy^+ + [1 + Strength(very)]$

because it is the dominant polarity. Similarly, in the second row negative polarity of 'bad' is increased.

Algorithm 2 Extended Classifier

INPUT: S,   Sentiment lexicon
                IntList,   Intensifier list
                DimList, Diminisher list
                NegList, Negation list
                Doc,   Document to be classified
OUTPUT: Class,   Sentiment class for d

1: **Initialise** Doc^+; Doc^-
2: **for all** $Sentence_k \in$ Doc **do**
3:     **for all** $t_i \in Sentence_k$ **do**
4:         **if** t_i is inflected for emphasis **then**
5:             convert t_i to standard form
6:             inflectionFlag $\leftarrow$ TRUE
7:         **end if**
8:         Retrieve t_i^+ and t_i^- from S
9:         **for all** $t_j \in (max(0, j - 5) \leq j \leq min(length(Sentence_i), j + 6))$ **do**
10:            **if** $j = i$ OR $t_j \in$ IntList OR $t_j \in$ DimList **then** next j
11:            **end if**
12:            **if** $t_j \in$ NegList **then** NegationShift(t_i^+, t_i^-)
13:            **else if** $t_j \in$ IntList **then** increment $max(t_i^+, t_i^-)$ by $Strength(t_j)$
14:            **else if** $t_j \in$ DimList **then** decrement $max(t_i^+, t_i^-)$ by $Strength(t_j)$
15:            **end if**
16:         **end for**
17:         **if** InflectionFlag=TRUE **then** increment $max(t_i^+, t_i^-)$ by 50%
18:         **end if**
19:         Normalize t_i^+ and t_i^-
20:         **if** $t_i^+ + t_i^- > 0$ **then**
21:            $Doc^+ \leftarrow Doc^+ + t_i^+$; $Doc^- \leftarrow Doc^- + t_i^+$
22:         **end if**
23:     **end for**
24: **end for**
25: Normalize Doc^+ and Doc^-
26: **if** $Doc^+ \geq Doc^-$ **then**
27:     **Return** *Positive*
28: **else**
29:     **Return** *Negative*
30: **end if**

---

5 Global Context

When Algorithm 2 is supplied with SentiWordNet as the sentiment lexicon, only sentiment-bearing terms that have an entry in SentiWordNet contribute towards to analysis. This means that potentially many domain-specific terms are likely to be ignored. Similarly some terms might have their sentiment context misrepresented in the lexicon as it only captures general purpose usage of terms. To address this limitation, we introduce a strategy to hybridize SentiWordNet with terms and sentiment context extracted from the domain of application.

The process of generating the hybrid lexicon is shown in Fig. 3. First, a domain-focused lexicon is generated from data extracted from the target domain and labelled using distant-supervision approach. Next the hybrid lexicon is generated by combining the sentiment scores (learnt for domain terms) in the domain-focused lexicon with existing scores in SentiWordNet. Next we look at each of these in turn.

5.1 Data Labelling: Distant Supervision

Distant-supervision offers an automated approach to assigning sentiment class labels to documents. It uses the presence of class-specific emoticons in a document as evidence for its true class. For example a smiley-face emoticon would according to distant-supervision be considered to be expressing positive sentiment and as such evidence for labelling the related content as belonging to the positive class. Accordingly, given a dataset and a lexicon of class-specific emoticons, we can assign such noisy labels to all documents that contain them in order to generate a labelled dataset for supervised learning tasks. In order to minimise the level of potential noise, a reasonable strategy is needed to process documents containing emoticons from both positive and negative classes. In this work, we avoid documents with mixed emoticons. We generate three distant-supervised datasets to represent varying web social communication settings (see Table 2) from blog messages (Digg and MySpace samples made available by cyberemotions.eu) to micro-blogs (twitter sample made available by

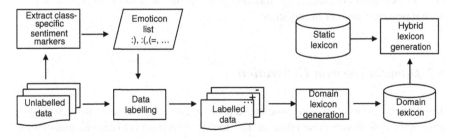

Fig. 3 Diagram showing stages involved in the proposed approach

Table 2 Datasets and sizes

Dataset	#Pos.	#Neg.	Avg. Sent. per Doc.	Avg. Character per Sent.	Description
Twitter	10,000	10,000	1.96	8.84	Variety of topics discussed, in mostly disconnected messages with a maximum of 140 characters
Digg	5222	5222	6.69	29.71	Messages from threaded discussions about various topics, with no imposed character limit per post
MySpace	292	292	4.36	5.80	Message exchanges between pair of Internet 'friends', with no imposed character limit per post. Tends to be mostly positive

sentiment140.com). With Digg and MySpace we restrict the labelling to sentences rather than documents as emoticons typically affect only the sentence in which they appear. This means that multiple micro-documents are generated from a single document ensuring each micro-document is labelled according to one or more emoticons belonging to the same sentiment class. With both these datasets there were many more positive (almost 80 %) compared to negative emoticons present. Accordingly a balanced sample from this extremely skewed distribution was used to create the distant-supervised datasets. The main difference between the Digg and MySpace generated datasets is in its size (Digg with 5222 and MySpace with 292 negative messages). Twitter, unlike with the other two, contained over a million distant-supervised tweets. We sampled 10,000 from each class (positive and negative) to generate a suitably sized dataset for our work, as the full dataset is too large to conveniently work with and the selected number is adequate to demonstrate our approach. All distance-supervised datasets are preprocessed to a reduced feature space using the approach introduced in [9]. That is, all user names (i.e. words that starts with the @ symbol) are replaced with the token USERNAME and URLs (e.g. http://tinyurl.com/cvvg9a) are replaced with the token URL. Moreover, words consisting of sequence of three or more repeated character (e.g. haaaaapy) are normalised to contain only two of such repeated character in sequence.

5.2 Domain Lexicon Generation

The domain-focused lexicon associates a positive and negative score to each unique term in the distant supervised dataset. Key to this generation is to capture association of a term t_i to a class c_j given a set of distant-supervised documents, D. To achieve this, we use a metric based on term frequency (TF). TF is the number of times a specific term appears in a document. It is a well established quantifier of association between documents in many text analysis tasks (e.g. Information Retrieval). We propose supervised TF (sTF) to associate terms with sentiment classes (positive and negative). sTF measures strength of association of a term t with a class c as the ratio of frequency of t_i in documents labelled as class c_j to the total frequency of t in all documents. This is shown in Eq. 1.

$$ds(t_i, c_j) = \frac{\text{TF}(t_i, D_c)}{\text{TF}(t_i, D)} \tag{1}$$

where, D_c is the subset of D labelled as class c, $\text{TF}(t_i, D)$ is the term frequency of t_i in D and $ds(t_i, c_j)$ is the domain-focused association of t_i with c_j.

5.3 Hybrid Lexicon Generation

Scores from SentiWordNet, S, and domain-focused, D, lexicon for each term t_i are combined to form the hybrid score for term (see Algorithm 3). When t_i appears in both lexicons, a weighted average of positive (and negative) scores supplied by both lexicons is calculated using α as a mixing parameter. So $\alpha = 0.5$ would lead to equal weighting of scores from S and D whilst $\alpha = 0$ will ignore scores from SentiWordNet lexicon (see steps 3 and 4). When only one lexicon (SentiWordNet or domain-focused) contains a scores for t_i, that score is fully used without an aggregation (see steps 6 and 8). Thereafter, the new scores for t_i (i.e. t_i^+ and t_i^-) are added to the hybrid lexicon, H (step 11). Finally, H is returned as the output.

6 Evaluation

We conduct a comparative study to evaluate the role of accounting for both local and global contexts of terms, as proposed in this work, for sentiment classification of social media text. The main aim of this study is to evaluate performances of the introduced strategies to account for local and global contexts individually and in combination. Also, we carry out further experiments to investigate the following.

- The contributions of individual lexicons (SentiWordNet and domain-focused) in the hybrid lexicon;
- The performance of the hybrid lexicon compared to machine learning approaches as distant-supervision is typically employed in machine learning; and
- The transferability of our approach across social media domains.

Accordingly we compare the following six strategies:

Algorithm 3 Generate Hybrid Lexicon

INPUT: S, Static lexicon
          D, Domain-focused Lexicon
          α, Unifying weight
OUTPUT: H, Hybrid lexicon
1: **for all** $t_i \in (S \cup D)$ **do**
2:     **if** $t_i \in S \cap D$ **then**
3:         $t_i^+ \leftarrow \alpha \times (t_i^+ \in S) + (1 - \alpha) \times (t_i^+ \in D)$
4:         $t_i^- \leftarrow \alpha \times (t_i^- \in S) + (1 - \alpha) \times (t_i^- \in D)$
5:     **else if** $t_i \in S$ **then**
6:         $t_i^+ \leftarrow (t_i^+ \in S)$
7:         $t_i^- \leftarrow (t_i^- \in S)$
8:     **else**
9:         $t_i^+ \leftarrow (t_i^+ \in D)$
10:       $t_i^- \leftarrow (t_i^- \in D)$
11:    **end if**
12:    $H.AddEntry(t_i^+, t_i^-)$
13: **end for**
14: **Return** H

---

1. BASE: Basic sentiment classification algorithm using SentiWordNet (i.e. Algorithm 1, using SentiWordNet)
2. DOMAIN: Basic sentiment classification algorithm using domain-focused lexicon (i.e. Algorithm 1, using domain-focused lexicon)
3. BASE+LC: An extension of the BASE algorithm with accounting for local context (i.e. Algorithm 2, using SentiWordNet)
4. BASE+GC: An extension of the BASE algorithm with accounting for global context (i.e. Algorithm 1, using hybrid lexicon)
5. BASE+LC+GC: An extension of the BASE algorithm with accounting for local and global contexts (i.e. Algorithm 2, using hybrid lexicon)
6. Machine Learning algorithms: We use three commonly used sentiment classification algorithms: Support Vector Machines (SVM), Naïve Bayes (NB) and Logistic Regression (LR).

All algorithms are tested using human labelled datasets from the three social media platforms, introduced earlier containing: 182 positive and 177 negative Twitter; 107 positive and 221 negative Digg; and 400 positive and 105 negative MySpace examples. The three machine learning classifiers are trained with the distant-supervised dataset. With all the algorithms, we employ binary feature representation of documents. A preliminary experiment with different values of α for the hybrid lexicon generation in Sect. 5.3 does not show a clear preference. Therefore, we use $\alpha = 0.5$ in this work thereby giving equal weights to both lexicons. As is typical with unbalanced datasets [13, 29] we present results based on the average value of the F1-score for positive and negative classes to quantify classification quality. Class-based precision (P) and recall (R) are also reported.

6.1 Local and Global Context

Tables 3, 4 and 5 show sentiment classification results on Twitter, Digg and MySpace datasets respectively with best results shown in bold. Combining Local and global contexts (BASE+LC+GC) performs best on Twitter and Digg datasets. Likewise, both BASE+LC and BASE+GC significantly improve upon BASE on these datasets. BASE+LC performs better than BASE+GC on Digg while BASE+GC performs better than BASE+LC on Twitter. This can be attributed to the fact that lexical modifiers, which BASE+LC accounts for, are more likely to appear in Digg than in Twitter due to the short length nature of tweets. Also, the number of distant-supervised tweets, which is about double the size of Digg, makes it more likely for the tweets classification to benefit more from hybrid lexicon (i.e. global context). This is more clear on the

Table 3 Performance of algorithms on Twitter dataset

Algorithm	Positive			Negative			Avg F1
	P	R	F1	P	R	F1	
Lexicon-based							
BASE	75.1	57.9	65.4	43.8	63.1	51.8	58.5
DOMAIN	65.2	71.5	68.2	73.2	67.2	70.1	69.2
BASE+LC	76.0	69.8	72.8	71.4	77.4	74.3	73.5
BASE+GC	71.8	76.0	73.9	76.7	72.6	74.6	74.2
BASE+LC+GC	85.5	72.7	77.7	67.8	80.0	73.4	**75.6**
Machine learning							
SVM	64.4	30.2	40.3	52.6	79.7	63.4	54.6
NB	62.6	64.3	63.4	62.2	60.5	61.3	62.4
LR	71.7	78.0	74.7	75.2	68.4	71.6	73.3

Table 4 Performance of algorithms on Digg dataset

Algorithm	Positive			Negative			Avg F1
	P	R	F1	P	R	F1	
Lexicon-based							
BASE	78.3	44.4	56.7	52.7	83.5	64.6	60.6
DOMAIN	78.5	41.6	54.4	45.3	81.2	58.2	56.3
BASE+LC	51.0	65.3	57.3	78.3	68.1	72.8	65.1
BASE+GC	84.1	46.0	59.5	51.2	87.1	64.5	62.0
BASE+LC+GC	93.5	50.0	65.2	54.8	94.5	69.4	**67.3**
Machine learning							
SVM	32.1	46.7	38.0	66.9	52.0	58.5	48.3
NB	32.3	46.7	38.2	67.1	52.5	58.9	48.6
LR	42.8	69.2	52.9	78.7	55.2	64.9	58.7

Table 5 Performance of algorithms on MySpace dataset

Algorithm	Positive			Negative			Avg F1
	P	R	F1	P	R	F1	
Lexicon-based							
BASE	79.9	86.7	83.2	52.9	40.7	46.0	64.6
DOMAIN	45.4	85.8	59.3	71.2	25.3	37.4	48.4
BASE+LC	91.5	78.3	84.4	46.6	72.4	56.7	**70.5**
BASE+GC	51.9	88.5	65.4	74.0	28.6	41.3	53.3
BASE+LC+GC	77.5	86.1	81.6	52.4	37.9	44.0	62.8
Machine learning							
SVM	79.2	100	88.4	0.0	0.0	0.0	44.2
NB	83.9	40.3	54.4	23.6	70.5	35.4	44.9
LR	88.0	67.8	76.6	34.5	64.8	45.0	60.8

much smaller MySpace dataset, where neither BASE+GC nor BASE+LC+GC perform better than BASE. Overall, the BASE+LC+GC approach performs better than all supervised machine learning algorithms (SVM, NB and LR) on all three datasets; 74.2 % versus 73.3 % on Twitter, 62.2 % versus 58.7 % on Digg and 62.9 % versus 60.8 % on MySpace. This confirms the superiority of our lexicon-based approach using a hybrid lexicon distant-supervised learning over the machine learning approaches to sentiment classification. The comparison of the lexicon based approaches to sentiment analysis shows that the hybrid lexicon does perform significantly better than the others with promising transferability prospects. These are discussed in more detail next.

6.2 Hybrid Versus Individual Lexicons

As expected, on Twitter dataset using the hybrid lexicon, BASE+GC, performs better than using either SentiWordNet, BASE, or domain-focused lexicon, DOMAIN, individually (74.2 % vs. 58.5 % and 69.2 % respectively). Also, DOMAIN performs better than BASE, indicating the inability of the static lexicon, which is generated from fairly standard text, to capture certain sentiment expressions from non-standard text. Similar results are also observed on Digg dataset. However, although best results are obtained with the hybrid lexicon, the BASE lexicon has out performed the DOMAIN lexicon. Although this difference is marginal it does raise two interesting questions: either distant-supervised labelling is more suitable for Twitter than Digg or the smaller distant-supervised data size in Digg, compared to Twitter, has affected the reliability of the domain-focused lexicon generated from Digg. It is also interesting to note that unlike on the Twitter dataset, all machine learning algorithms have performed extremely poorly on the Digg dataset. Given that they rely heavily on the distant-supervised labelled data (just as the DOMAIN algorithms) it is likely that considerable noise has been introduced by relying on sentiment markers from a poorly representative sample of data.

In fact examining results from MySpace (the smallest of the three datasets for distant-supervision) further supports this observation. Once again we see poor accuracy with machine learning algorithms and BASE performing better than DOMAIN and comparable to BASE+GC. This is more likely to be caused by the very limited data from which the a domain-focused lexicon is generated for MySpace. This suggests the need to establish minimum dataset requirement below which a domain-focused lexicon becomes unreliable due to atypical usage of emoticons such as when used to express sarcasm or to soften intensity of their opposite sentiment. This then begs the question of can we augment smaller distant-supervised datasets that are likely to be less representative of the underlying emoticon usage behaviour with larger datasets that are easier to obtain from a different domain. This issue brings us conveniently onto the next topic of transferability.

6.3 Transferability Across Social Media Platforms

As distant-supervision relies on certain sentiment markers to label documents which may not be very common in some social media platforms, it is imperative to asses performance of the hybrid lexicon on a platform different from the one it was initially generated on (i.e. transferability of the lexicon). We investigate behaviour of BASE+GC when distant-supervised data is generated from a different social media platform. Results are shown in Table 6, plus sign (+) indicates improvement over using within domain lexicon (BASE+GC) while the minus sign (−) indicates a decline. For Twitter, using its own domain for distant-supervision (i.e. within platform) is better than either using Digg posts or MySpace messages (74.2 vs. 62.1 and 60.6). However with the other smaller distant-supervised datasets (Digg and MySpace) we see significant improvements when they are augmented or replaced with the larger Twitter distant-supervised dataset. For instance with Digg an increase of over 5 % is observed, when using a distant-supervised Twitter dataset. Whilst with MySpace an impressive 10 % improvement is observed with a distant-supervised dataset formed by combining all platforms. These results indicate that where within platform dataset is small or unavailable, using data from a different platform is advantageous. However, the results on MySpace show that there is still the question of platform compatibility, the Digg generated lexicon compares favourably over the Twitter lexicon even though the size of the distant-supervised Twitter dataset is a magnitude larger than Digg dataset. A drill down into the precision and recall values reveal that both SentiWordNet and MySpace lexicons have the same result pattern: high precision and low recall on both positive and negative classes. Therefore, they tend to have uncomplementary strength. Digg lexicon, however, shows a low precision and high recall on both positive and negative classes, thus, has complementary strength with SentiWordNet.

Table 6 Results of transferability of hybrid lexicon across social media platforms

Source/Test platforms	Positive			Negative			Avg F1
	P	R	F1	P	R	F1	
Twitter as distant-supervised dataset							
Digg	67.9	55.8	61.2	74.1	82.7	78.2	69.7+
MySpace	60.4	90.6	72.5	76.0	33.3	46.3	59.4−
Digg as distant-supervised dataset							
MySpace	83.2	87.4	85.2	53.8	45.5	49.3	67.3+
Twitter	71.3	61.1	65.8	53.4	64.4	58.4	62.1−
MySpace as distant-supervised dataset							
Twitter	43.1	70.3	53.4	81.3	58.1	67.8	60.6−
Digg	41.5	52.4	46.3	81.8	74.4	77.9	62.1+
All platforms as source							
Twitter	70.4	73.1	71.7	73.4	70.8	72.1	71.9−
Digg	67.4	70.1	68.7	70.4	67.6	69.1	68.3+
MySpace	87.4	90.0	85.7	65.4	48.2	55.5	70.6+

7 Conclusions and Future Work

In this work, we presented an approach to addressing the semantic gap between prior and contextual polarities of terms for opinion mining tasks. We confirm previous research that confirm the usefulness of local context in specific, negation, intensifiers, diminishers and other non lexical modifiers. Another aspect of the semantic gap is the difference in vocabulary coverage and term usage between a lexicon and its domain of application. We presented a novel approach capture this global context through the generation of a hybrid lexicon that enhances a general purpose lexicon (SentiWordNet) with domain knowledge for sentiment classification. We demonstrated how distant-supervision can be exploited for this purpose. Experimental evaluation shows that the approach is effective and better than state-of-the-art machine learning sentiment classification trained on the same dataset from which domain knowledge is extracted in our approach (i.e. distant-supervised data). Combining all distant-supervised data from the three domains leads to an overall significant performance improvements with the hybrid lexicon, confirming the transferability of the lexicon across social media platforms. This also suggests that combining distant-supervised data from multiple social media platforms may help especially where there is no sufficient data from a target platform. However we also observed that there are compatibility issues between domains that warrants further investigation. In future work we will explore how characterising a dataset might help towards addressing this issue.

References

1. Arnold, I., Vrugt, E.: Fundamental uncertainty and stock market volatility. Appl. Financ. Econ. 18(17), 1425–1440 (2008)
2. Baccianella, S., Esuli, A., Sebastiani, F.: Sentiwordnet 3.0: an enhanced lexical resource for sentiment analysis and opinion mining. In: Proceedings of the Annual Conference on Language Resouces and Evaluation (2010)
3. Baron, D.: Competing for the public through the news media. J. Econ. Manag. Strategy 14(2), 339–376 (2005)
4. Brin, S., Page, L.: The anatomy of a large-scale hypertextual web search engine. In: Seventh International World-Wide Web Conference (WWW 1998) (1998)
5. Dang, Y., Zhang, Y., Chen, H.: A lexicon-enhanced method for sentiment classification: an experiment on online product reviews. IEEE Intell. Syst. 25, 46–53 (2010)
6. Denecke, K.: Using sentiwordnet for multilingual sentiment analysis. In: ICDE Workshop (2008)
7. Esuli, A., Baccianella, S., Sebastiani, F.: Sentiwordnet 3.0: an enhanced lexical resource for sentiment analysis and opinion mining. In: Proceedings of the Seventh Conference on International Language Resources and Evaluation (LREC10) (2010)
8. Fellbaum, C.: WordNet: An Electronic Lexical Database. MIT Press, Cambridge (1998)
9. Go, A., Bhayani, R., Huang, L.: Twitter sentiment classification using distant supervision. Processing pp. 1–6 (2009)
10. Hu, M., Liu, B.: Mining and summarizing customer reviews. In: Proceedings of the Tenth ACM SIGKDD International Conference on Knowledge Discovery and Data Mining, pp. 168–177 (2004)
11. Karlgren, J., Sahlgren, M., Olsson, F., Espinoza, F., Hamfors, O.: Usefulness of sentiment analysis. In: 34th European Conference on Information Retrieval (2012)
12. Kennedy, A., Inkpen, D.: Sentiment classification of movie reviews using contextual valence shifters. Comput. Intell. 22, 2006 (2006)
13. Li, Y., Bontcheva, K., Cunningham, H.: Using uneven margin svm and perceptron for information extraction. In: Proceedings of the Conference on Natural Language Learning (CONLL'05), pp. 72–79 (2005)
14. Liu, B.: Sentiment Analysis and Subjectivity, 2nd edn., Chapter Handbook of Natural Language Processing, pp. 627–666. Chapman and Francis, Boca Raton (2010)
15. Ludvigson, S.: Consumer confidence and consumer spending. J. Econ. Perspect. 18(2), 29–50 (2004)
16. Mohammad, S.M., Kiritchenko, S., Zhu, X.: Nrc-canada: building the state-of-the-art in sentiment analysis of tweets. In: Proceedings of the Seventh International Workshop on Semantic Evaluation Exercises (SemEval-2013). Atlanta, Georgia (2013)
17. Muhammad, A., Wiratunga, N., Lothian, R., Glassey, R.: Contextual sentiment analysis in social media using high-coverage lexicon. In: Research and Development in Intelligent Systems XXX, pp. 79–93. Springer, New York (2013)
18. Ohana, B., Tierney, B.: Sentiment classification of reviews using sentiwordnet. In: 9th IT&T Conference, Dublin, Ireland (2009)
19. Paltoglou, G., Thelwall, M.: Twitter, myspace, digg: unsupervised sentiment analysis in social media. ACM Trans. Intell. Syst. Technol. 3(4), (2012)
20. Pang, B., Lee, L.: Opinion mining and sentiment analysis. Found. Trends Inf. Retrieval 2(1), 1–135 (2008)
21. Pang, B., Lee, L., Vaithyanathan, S.: Thumbs up? Sentiment classification using machine learning techniques. In: Proceedings of the Conference on Empirical Methods on Natural Language Processing (2002)
22. Pera, M., Qumsiyeh, R., Ng, Y.K.: An unsupervised sentiment classifier on summarized or full reviews. In: Proceedings of the 11th International Conference on Web Information Systems Engineering, pp. 142–156 (2010)

23. Polanyi, L., Zaenen, A.: Contextual Valence Shifters, vol. 20. Springer, Dordrecht, The Netherlands (2004)
24. Read, J.: Using emoticons to reduce dependency in machine learning techniques for sentiment classification. In: Proceedings of the ACL Student Research Workshop. ACLstudent '05, pp. 43–48. Association for Computational Linguistics, Stroudsburg (2005)
25. Riloff, E., Patwardhan, S., Wiebe, J.: Feature subsumption for opinion analysis. In: Proceedings of the 2006 Conference on Empirical Methods in Natural Language Processing (EMNLP-06) (2006)
26. Stone, P.J., Dexter, D.C., Marshall, S.S., Daniel, O.M.: The General Inquirer: A Computer Approach to Content Analysis. MIT Press, Cambridge (1966)
27. Taboada, M., Brooke, J., Tofiloski, M., Voll, K., Stede, M.: Lexicon-based methods for sentiment analysis. Comput. Linguist. **37**, 267–307 (2011)
28. Thelwall, M., Buckley, K., Paltoglou, G.: Sentiment strength detection for the social web. J. Am. Soc. Inf. Sci. Technol. **63**(1), 163–173 (2012)
29. Thelwall, M., Buckley, K., Paltoglou, G., Cai, D., Kappas, A.: Sentiment strength detection in short informal text. J. Am. Soc. Inf. Sci. Technol. **61**(12), 2444–2558 (2010)
30. Whitelaw, C., Garg, N., Argamon., S.: Using appraisal groups for sentiment analysis. In: 14th ACM International Conference on Information and Knowledge Management (CIKM 2005), pp. 625–631 (2005)

Case-Studies in Mining User-Generated Reviews for Recommendation

Ruihai Dong, Michael P. O'Mahony, Kevin McCarthy and Barry Smyth

Abstract User-generated reviews are now plentiful online and they have proven to be a valuable source of real user opinions and real user experiences. In this chapter we consider recent work that seeks to extract topics, opinions, and sentiment from review text that is unstructured and often noisy. We describe and evaluate a number of practical case-studies for how such information can be used in an information filtering and recommendation context, from filtering helpful reviews to recommending useful products.

1 Introduction

User-generated reviews are now a common feature of online sites and stores. They have proven to be an important source of user opinions on products and services from books and movies to accommodation, people, and electronics. In fact, user-generated reviews and now considered by many to be a vital part of how users inform themselves, especially when it comes to purchasing behaviour. It is largely accepted that availability of reviews helps shoppers to choose [1] and increases the likelihood that they will make a buying decision [2], for example.

In this chapter we are interesting in automatically mining valuable opinion information from this plentiful but unstructured, and often noisy, source of user

R. Dong (✉)
CLARITY: Centre for Sensor Web Technologies, University College Dublin,
Dublin, Ireland
e-mail: ruihai.dong@ucd.ie

M.P. O'Mahony · K. McCarthy · B. Smyth
Insight Centre for Data Analytics, University College Dublin, Dublin, Ireland
e-mail: michael.omahony@insight-centre.org

K. McCarthy
e-mail: kevin.mccarthy@insight-centre.org

B. Smyth
e-mail: barry.smyth@insight-centre.org

© Springer International Publishing Switzerland 2015   105
M.M. Gaber et al. (eds.), *Advances in Social Media Analysis*,
Studies in Computational Intelligence 602,
DOI 10.1007/978-3-319-18458-6_6

knowledge. We do this primarily by using shallow natural language processing, topic mining, and sentiment analysis techniques and demonstrate how the resulting information can be applied in a variety of information filtering and recommendation tasks. We begin by reviewing a representative sample of the state of the art in opinion mining with a particular focus on user-generated reviews. Next we describe our approach to topic extraction and sentiment analysis that is at the heart of our opinion mining method. We describe a series of case-studies to demonstrate some of the ways that the topics and opinions extracted from user-generated reviews can be applied in practice. For example, in the first case-study we look at the familiar task of classifying reviews and predicting review helpfulness [3–10] to demonstrate how an opinion-mining approach can offer some advantage over conventional alternatives. Following on, our second case-study describes a straight forward technique for recommending informative reviews to users based on our ability to accurately predict review helpfulness. Finally, in our third case-study we move from dealing with single reviews to using a collection of product reviews as a new source of product information. We describe and evaluate a product recommender that harnesses product descriptions that are formed exclusively from the opinions found in user reviews and show how this approach provides a novel basis to generate recommendations.

2 Related Work

Recent research highlights how online product reviews can influence the purchasing behaviour of users; see [1, 2]. The effect of consumer reviews on book sales on Amazon.com and Barnesandnoble.com [11] shows that the relative sales of books on a site correlates closely with positive review sentiment; although interestingly, there was insufficient evidence to conclude that retailers themselves benefit from making product reviews directly available to consumers; see also the work of [12, 13] for music and movie sales, respectively. As a result researchers have begun to focus on harnessing this type of user-generated content and there are two areas of related work particularly relevant to the research presented in this chapter: classifying reviews and extracting opinions from reviews.

2.1 Classifying User-Generated Reviews

As review volume has grown retailers recognise the need to develop ways to help users find high quality reviews for products of interest and to avoid malicious or biased reviews. This has led to a body of research focused on classifying or predicting review helpfulness, and also research on detecting so-called *spam reviews.*

Review helpfulness classification approaches, such as that proposed in [3], typically consider features related to the ratings, structural, syntactic, and semantic properties of reviews and have found ratings and review length among the most discriminating. Reviewer expertise was found to be a useful predictor of review helpfulness in [4], confirming, in this case, the intuition that people interested in a certain genre of movies are likely to pen high quality reviews for similar genre movies. Review timeliness was also found to be important since review helpfulness declined as time went by. Furthermore, opinion sentiment has been mined from user reviews to predict ratings and helpfulness in services such as TripAdvisor [5–8].

Just as it is useful to automate the filtering of helpful reviews it is also important to identify malicious or biased reviews. These reviews can be well written and informative and so can appear to be helpful. However these reviews often adopt a biased perspective that is designed to help or hinder sales of the target product [9]. Li et al. [10] describe a machine learning approach to spam detection that is enhanced by information about the spammer's identity as part of a two-tier, co-learning approach. On a related topic, network analysis techniques are used in [14] to identify recurring spam in user-generated comments associated with YouTube videos by identifying discriminating comment *motifs* that are indicative of spambots.

2.2 Mining Opinions and Features from User-Generated Reviews

There have also been a number of efforts focused on the extraction of feature-based product descriptions from user reviews. The work in [15] is representative in this regard and describes the use of shallow natural language processing (NLP) techniques for explicit feature extraction and sentiment analysis; see also [16, 17]. The features extracted, and the techniques used, are similar to those presented in this chapter, although in the case of the former there was a particular focus on the extraction of merenomic and taxonomic features to describe the *parts* and *properties* of a product. In [18], the sentiment of comparative and subjective sentences in reviews is analysed on a per-feature basis to create a semi-order of products, but the recommendation task with respect to a query product is not considered.

In this chapter we are particularly interested in product recommendation and the ability of review opinions to inform the recommendation process. Conventional recommender systems are either based on ratings or transaction data (collaborative filtering) or on fixed content representations (content-based filtering), and the idea of developing a recommendation framework based purely on noisy user-generated content remains novel in itself. The work in [19] is relevant in this regard in that it uses user-generated micro reviews as the basis for a text-based content recommender, and recently work in [20] has also tried to exploit user-generated content in similar ways. Likewise, reviews are leveraged to alleviate the well-known cold-start problem associated with collaborative recommenders [21]. In this work, the focus is on mining user preferences from review texts to reduce the sparsity of the user-item matrix; thereafter standard collaborative filtering algorithms are applied to the augmented user-item matrix to improve recommendation performance.

3 Topic Extraction and Sentiment Analysis from User-Generated Reviews

The main focus of this chapter is how topics and sentiment mined from user-generated product reviews can be leveraged as the basis for new approaches to product filtering and recommendation. Before we describe how this topical and sentiment information can be used in practice, the approach to automatically extract topics and assign sentiment is first described; see Fig. 1 for an overview of this approach.

It is worth highlighting that the approach uses a combination of existing techniques from the literature; no novel techniques are presented. Rather, the main interest lies is the novel ways in which the extracted information can be applied to the filtering and recommendation of products as described in the case-studies that follow.

3.1 Topic Extraction

We consider two basic types of topics—*bi-grams* and *single nouns*—which are extracted using a combination of shallow NLP and statistical methods, primarily by combining ideas from [16, 22]. In the pre-processing step, we use OpenNLP[1] to

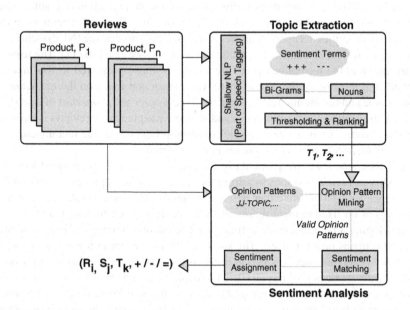

Fig. 1 System architecture for extracting topics and associated sentiment from user generated reviews

---

[1]OpenNLP: http://incubator.apache.org/opennlp/.

split reviews into sentences and label each term in a sentence with its appropriate part of speech, such as *NNS* (Noun, plural), *JJ* (Adjective), *VB* (Verb, base form) etc. Then, all terms in sentences are converted to lowercase and stemmed to root form, and in our method, stop words are excluded. To produce a set of bi-gram topics, all bi-grams from the global sentence set are extracted which conform to one of two basic part-of-speech co-location patterns: (1) an adjective followed by a noun (AN), such as *wide angle*; and (2) a noun followed by a noun (NN), such as *video mode*. These are candidate topics that need to be filtered to avoid including AN's that are actually opinionated single-noun topics; for example, *excellent lens* is a single-noun topic (*lens*) and not a bi-gram topic. Thus, bi-grams whose adjective is found to be a sentiment word (e.g. *excellent, good, great, lovely, terrible, horrible* etc.) are excluded using the sentiment lexicon proposed in [17].

To identify the single-noun topics we extract a candidate set of (non stop-word) nouns from the global review set. Often these single-noun candidates will not make for good topics; for example, they might include words such as *family* or *day* or *vacation*. A solution for validating such topics is proposed in [23] by eliminating those that are rarely associated with opinionated words. The intuition is that nouns that frequently occur in reviews and that are frequently associated with sentiment rich, opinion laden words are likely to be product topics that the reviewer is writing about, and therefore represent valid topics. Thus, for each candidate single-noun, how frequently it appears with nearby words from a list of sentiment words (using Hu and Liu's sentiment lexicon as above) is calculated, keeping the single-noun only if this frequency is greater than some threshold (in this case 30%).

The result is a set of bi-gram and single-noun topics which is further filtered based on their frequency of occurrence in the review set, keeping only those topics $(T_1, \ldots, T_m)$ that occur in at least k reviews out of the total number of n reviews; by experiment, $k_{bg} = n/20$ is used for bi-gram topics and $k_{sn} = 10 \times k_{bg}$ for single noun topics.

3.2 Sentiment Analysis

To determine the sentiment of the topics in the product topic set, a method similar to the *opinion pattern mining* technique [24] is used for extracting opinions from unstructured product reviews. Once again the sentiment lexicon from [17] is used as the basis for this analysis. For a given topic T_i, and corresponding review sentence S_j from review R_k (that is the sentence in R_k that includes T_i), any sentiment words in S_j are identified. If there are none then this topic is marked as *neutral* from a sentiment perspective. If sentiment words $(w_1, w_2, \ldots)$ are present, that sentiment word (w_{min}) which has the minimum word-distance to T_i is identified.

Next the part-of-speech tags for w_{min}, T_i and any words that occur between w_{min} and T_i are determined. The POS sequence corresponds to an opinion pattern. For example, in the case of the bi-gram topic *noise reduction* and the review sentence

"...this camera has great noise reduction...", w_{min} is the word *"great"* which corresponds to the opinion pattern *JJ-TOPIC* as per [24].

Once an entire pass of all topics has been completed, the frequency of all opinion patterns that have been recorded is computed. A pattern is deemed to be valid (from the perspective of our ability to assign sentiment) if it occurs more than the average number of occurrences over all patterns [24]. For valid patterns sentiment is assigned based on the sentiment of w_{min} and subject to whether S_j contains any negation terms within a 4-word-distance^2 of w_{min}. If there are no such negation terms then the sentiment assigned to T_i in S_j is that of the sentiment word in the sentiment lexicon. If there is a negation word then this sentiment is reversed. If an opinion pattern is deemed not to be valid (based on its frequency) then a *neutral* sentiment is assigned to each of its occurrences within the review set.

4 Case-Study 1: Predicting Review Helpfulness

In the previous section an approach to automatically mine topics $(T_1, \ldots, T_m)$ and associated sentiment from review texts was described. Thus, each review R_i can be associated with *sentiment tuples*, $(R_i, S_j, T_k, +/-/ =)$, corresponding to a sentence S_j containing topic T_k with a sentiment value positive (+), negative (−), or neutral (=). This approach forms the basis of a number of case-studies to explore how to harness user-generated reviews in various recommendation and recommendation-related tasks. To begin, in this first case-study, the task of classifying helpful reviews is examined, based on a variety of classification features, including the topical and sentiment features described above. The key question that will be explored is whether these topical and sentiment features add value relative to traditional features used in review classification.

4.1 Classifying Helpful Reviews

To build a classifier for predicting review helpfulness, a supervised machine learning approach is adopted. In the data that is available to us each review has a helpfulness score that reflects the percentage of positive votes that it has received, if any. Following the approach described in [8], a review is labeled as *helpful* if and only if it has a helpfulness score in excess of 0.75. All other reviews are labeled as *unhelpful*.

To represent review instances, a standard feature-based encoding is used based on a set of 7 different types of features, including temporal information (AGE), rating

---

^{2}In long sentences, users may comment on multiple features. Thus, we introduce a window size for negation terms to limit their scope to nearby features. Based on experiment, we set the window size to four. Moreover, we identify certain phrases (e.g. "not only") which are not considered from a sentiment perspective. We acknowledge that more sophisticated sentiment analysis techniques have been proposed, an investigation of which we leave to future work.

information (RAT), simple sentence and word counts ($SIZE$), topical coverage (TOP), sentiment information ($SENT$), readability metrics ($READ$), and content made up of the top 50 most popular topics extracted from the reviews (CNT). These different types, and the corresponding individual features are summarised in Table 1. Some of these features, such as rating, word and sentence length, date and readability have been considered in previous work [3, 4, 26] and reflect best practice in the field of review classification. However, the topical and sentiment features (explained in detail below) are novel, and the comparison of the performance of the different feature sets is intended to demonstrate the efficacy of these new features, in isolation and combination, and in comparison to classical benchmarks across a common dataset and experimental configurations.

4.2 From Topics and Sentiment to Classification Features

As described above, a set of topics ($topics(R_k) = T_1, T_2, \ldots, T_m$) and corresponding sentiment scores (*pos/neg/neutral*) is assigned to each review R_k, which can be considered in isolation and/or in aggregate as the basis for classification features. For example, information about a review's *breadth* and *depth* of topic coverage can be obtained by simply counting the number of topics contained within the review and the average word count associated with the corresponding review sentences; see Eqs. 1 and 2. Similarly, the popularity of review topics, relative to the topics across the product as a whole, is given by Eq. 3, where $rank(T_i)$ is a topic's popularity rank for the product and $UniqueTopics(R_k)$ as the set of unique topics in a review. Thus, if a review covers many popular topics then it receives a higher $TopicRank$ score than if it covers fewer rarer topics.

$$Breadth(R_k) = |topics(R_k)| \qquad (1)$$

$$Depth(R_k) = \frac{\sum_{\forall T_i \in topics(R_k)} len(sentence(R_k, T_i))}{Breadth(R_k)} \qquad (2)$$

$$TopicRank(R_k) = \sum_{\forall T_i \in UniqueTopics(R_k)} \frac{1}{rank(T_i)} \qquad (3)$$

Regarding sentiment, a variety of classification features can be derived: the number of positive (*NumPos* and *NumUPos*), negative (*NumNeg* and *NumUNeg*) and neutral (*NumNeutral* and *NumUNeutral*) topics (total and unique) in a review; the rank-weighted number of positive (*WPos*), negative (*WNeg*), and neutral (*WNeutral*) topics; the relative sentiment, positive (*RelUPos*), negative (*RelUNeg*), or neutral (*RelUNeutral*), of a review's topics. These features are all summarised in Table 1 under $SENT$.

Table 1 Classification feature sets

Type	Feature	#	Description
AGE	Age	1	The number of days since the review was posted
RAT	$NormUserRating$	1	A normalised rating score obtained by scaling the user's rating into the interval [0, 1]
SIZE	$NumSentences$	1	The number of sentences in the review text
	$NumWords$	1	The total number of words in the review text
TOP	$Breadth$	1	The total number of topics mined from the review
	$Depth$	1	The average number of words per sentence containing a mined topic
	$Redundancy$	1	The total word-count of sentences that are not associated with any mined topic
	$TopicRank$	1	The sum of the reciprocal popularity ranks for the mined topics present; popularity ranks are calculated across the target product
SENT	$NumPos$ (Neg, $Neutral$)	3	The number of positive, negative, and neutral topics, respectively
	$Density$	1	The percentage of review topics associated with non-neutral sentiment
	$NumUPos$ (Neg, $Neutral$)	3	The number of *unique* topics with positive/negative/neutral sentiment
	$WPos$ (Neg, $Neutral$)	3	The number of positive, negative, and neutral topics, weighted by their reciprocal popularity rank
	$RelUPos$ (Neg, $Neutral$)	3	The relative proportion of unique positive/negative/neutral topics
	$SignedRatingDiff$	1	The value of $RelUPos$ minus $NormUserRating$
	$UnsignedRatingDiff$	1	The absolute value of $RelUPos$ minus $NormUserRating$
READ	$NumComplex$	1	The number of 'complex' words (3 or more syllables) in the review text
	$SyllablesPerWord$	1	The average number of syllables per word
	$WordsPerSen$	1	The average number of words per sentence
	$GunningFogIndex$	1	The number of years of formal education required to understand the review
	$FleschReadingEase$	1	A standard readability score on a scale from 1 (30—very difficult) to 100 (70—easy)
	$KincaidGradeLevel$	1	Translates FleschReadingEase into KincaidGradeLevel required (U.S. grade level)
	$SMOG$	1	Simple Measure of Gobbledygood (SMOG) estimates the years of education required, see [25]
CNT		50	The top 50 most frequent topics that occur in a particular product's reviews

Also considered is a measure of the relative *density* of opinionated (non-neutral sentiment) topics in a review (see Eq. 4) and a relative measure of the difference between the overall review sentiment and the user's normalized product rating, i.e. $SignedRatingDiff(R_k) = RelUPos(R_k) - NormUserRating(R_k)$; we also compute an unsigned version of this metric. The intuition behind the rating difference metrics is to note whether the user's overall rating is similar to or different from the positivity of their review content. Finally, as shown in Table 1, each review instance also encodes a vector of the top 50 most popular review topics (*CNT*), indicating whether it is present in the review or not.

$$Density(R_k) = \frac{|pos(topics(R_k))| + |neg(topics(R_k))|}{|topics(R_k)|} \qquad (4)$$

4.3 Expanding Basic Features

Each of the basic features in Table 1 is calculated for a particular review. For example, the *breath* of review R_k may be 5, indicating that it covers 5 identified topics. Whether this represents a high or low value for the product in question in unclear, which may have tens or even hundreds of reviews written about it. For this reason, in addition to this basic feature value, 4 other variations are calculated as follows to reflect the distribution of its values across a particular product:

- The *mean* value for this feature across the set of reviews for the target product.
- The *standard deviation* of the values for this feature across the target product reviews.
- The *normalised* value for the feature based on the number of standard deviations above (+) or below (−) the mean.
- The *rank* of the feature value, based on a descending ordering of the feature values for the target product.

Accordingly most of the features outlined in Table 1 translate into 5 different actual features (the original plus the 4 variations) for use during classification. This is the case for every feature (30 in all) in Table 1 except for the content features (*CNT*). Thus each review instance is represented as a total set of 200 features $((30 \times 5) + 50$ features).

4.4 Evaluation

Our hypothesis is that the topical and sentiment features will help when it comes to the automatic classification of user generated reviews, into *helpful* and *unhelpful* categories, by improving classification performance above and beyond more traditional features (e.g. terms, ratings, readability etc.); see [3, 7]. This hypothesis is tested on real-world review data for a variety of product categories using a number of different classifiers.

4.4.1 Datasets and Methodology

The review data for this experiment was extracted from Amazon.com during October 2012; in total, 51,837 reviews for 1,384 unique products were collected. Reviews for 4 product categories—*Digital Cameras (DC), GPS Devices, Laptops, Tablets*—were considered and each was labeled as *helpful* or *unhelpful*, depending on whether their helpfulness score was above 0.75 or not, as described in Sect. 4.1. For the purpose of this experiment, all reviews included at least 5 helpfulness scores (to provide a reliable ground-truth) and the helpful and unhelpful sets were sampled so as to contain approximately the same number of reviews. Table 2 presents a summary of these data, per product type, including the average helpfulness scores across all reviews, and separately for helpful and unhelpful reviews.

Each review was processed to extract the classification features as described above. Here we are particularly interested in understanding the classification performance of different categories of features. In this case, 8 different categories are considered, *AGE, RAT, SIZE, TOP, SENT-1, SENT-2, READ, CNT*. Note, the sentiment features (*SENT*) are subdivided into into two groups *SENT-1* and *SENT-2*. The latter contains all of the sentiment features from Table 1 whereas the former excludes the ratings difference features (signed and unsigned) so that the influence of rating information (usually a powerful classification feature in its own right) within the sentiment feature-set can be better understood. Accordingly, corresponding datasets for each category (Digital Cameras, GPS Devices, Laptops and Tablets) were created in which the reviews were represented by a single set of features; for example, the *SENT-1* dataset consists of reviews (one set of reviews for each product category) represented according to the *SENT-1* features only.

For the purpose of this evaluation three commonly used classifiers were considered: *RF (Random Forest), JRip* and *NB (Naïve Bayes)*, see [27]. In each case classification performance was evaluated in terms of the area under the ROC curve (AUC) using 10-fold cross validation.

Table 2 Filtered and balanced dataset statistics

Category	#Reviews	#Prod.	Avg. Helpfulness		
			Help.	Unhelp.	All
DC	3180	113	0.93	0.40	0.66
GPS Devices	2058	151	0.93	0.46	0.69
Laptops	4172	592	0.93	0.40	0.67
Tablets	6652	241	0.92	0.39	0.65

4.4.2 Results

The results are presented in Figs. 2, 3 and 5. In Figs. 2, 3 and 4 the AUC performance for each classification algorithm (RF, JRip, NB) is shown separately; each graph plots the AUC of one algorithm for the 8 different categories of classification features for each of the four different product categories (DC, GPS, Laptop, and Tablet). Figure 5 provides a direct comparison of all classification algorithms (RF, JRip, NB); here results for a classifier using all features combined are presented. AUC values in excess of 0.7 can be considered as useful from a classification performance viewpoint [28]. Overall it can be seen that RF tends to produce better classification performance across the various feature groups and product categories. Classification performance tends to be poorer for the GPS dataset compared to Laptop, Tablet, and DC.

Previous research indicates that ratings information proves to be particularly useful when it comes to evaluating review helpfulness; see [3]. It is not a surprise therefore to see our ratings-based features perform well, often achieving an AUC > 0.7 on their own. For example, in Fig. 2 an AUC of approximately 0.75 for the Laptop and Tablet datasets is achieved, compared to between 0.65 and 0.69 for GPS and DC, respectively. Other 'traditional' feature groups (AGE, SIZE, READ, and CNT) rarely achieve AUC scores > 0.7 across the product categories.

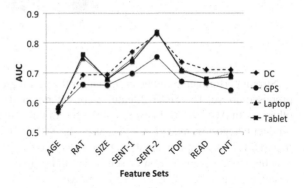

Fig. 2 Classification performance results for the RF classifier and different feature groups

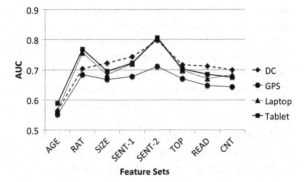

Fig. 3 Classification performance results for the JRip classifier and different feature groups

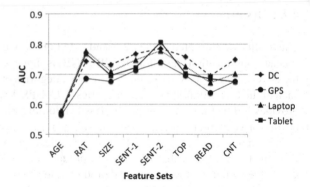

Fig. 4 Classification performance results for the NB classifier and different feature groups

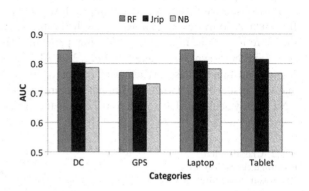

Fig. 5 Comparison of RF, JRip and NB for all features

Strong performance can be observed for the new topic and sentiment feature-sets proposed above. The *SENT-2* features consistently and significantly outperform all others, with AUC scores in excess of 0.7 for all three algorithms and across all four product categories; indeed in some cases the *SENT-2* features deliver AUC greater than 0.8 for DC, Laptop and Tablet products; see Fig. 2. The *SENT-2* feature group benefits from a combination of sentiment and ratings based features but a similar observation can be made for the sentiment-only features of *SENT-1*, which also achieve AUC greater than 0.7 for almost all classification algorithms and product categories. Likewise, the topical features (*TOP*) also deliver a strong performance with $AUC > 0.7$ for all product categories except for *GPS*.

These results bode well for a practical approach to review helpfulness prediction/ classification, with or without ratings data. The additional information contained within the topical and sentiment features contributes to an uplift in classification performance, particularly with respect to more conventional features that have been traditionally used for review classification. In Fig. 5, summary classification results according to product category are presented when classifiers are trained using a combination of all feature types. Once again strong classification performance is achieved; for example, an AUC of more than 0.7 for all conditions is achieved and the *RF* classifier delivers an AUC close to 0.8 or beyond for all categories.

5 Case-Study 2: Recommending Helpful Reviews

On many e-commerce sites users are faced with having to sift through hundreds or even thousands of reviews, depending on the popularity of products. In the previous case-study we demonstrated that it is possible to accurately predict whether a given review is likely to be helpful or not. Given the review overload facing users it is worthwhile to consider taking this approach a step further: instead of classifying the helpfulness of a single review, can a review or set of reviews be identified for recommendation to a user, given their interest in a specific product? Hence in this case-study an approach to turning our review classifier into a review recommender is described.

5.1 From Helpfulness Classification to Review Recommendation

Amazon currently adopts a simple approach to review recommendation, by suggesting the most helpful positive and most helpful critical review from a review collection. Amazon collects review helpfulness feedback to support this form of review recommendation and as a criterion to rank reviews. But this approach is far from perfect. Many reviews (often a majority) have received very few or no helpfulness ratings. This is especially true for more recent reviews, which arguably may be more reliable in the case of certain product categories (e.g. hotel rooms). Moreover, if reviews are ranked by helpfulness then it is unlikely that users will see those yet to be rated, making it even less likely that they will attract ratings. It quickly becomes a case of *"the rich get richer"* for those early-rated helpful reviews.

Therefore, the motivation for this case study is to examine, in the absence of review helpfulness information, whether it is possible to make useful review recommendations. In Sect. 4 it was shown that reviews can be accurately classified as helpful or not, but what about identifying the *most* helpful review or a set of the most helpful reviews for a given product? In what follows, this question is considered by showing how the review classifier can be used to recommend helpful reviews to a user. In particular, classification confidence is used as the basis for the recommendation ranking. Thus, for a given product, the rank order of a recommended review is given by the classification confidence that the review is helpful.

5.2 Evaluation

In this experiment, review data for the 4 product categories—Digital Cameras (DC), GPS Devices, Laptops, Tablets—as described in Sect. 4.4.1 are used. For each product category, a 10 fold cross validation experimental methodology was used, such that each review for each product was associated with a classification confidence that the review was helpful. The reviews for each product were then ranked by classification confidence and the top-ranked review was recommended; this approach is referred to as the *Pred* strategy. Recall this recommendation is made without the presence of actual helpfulness scores and relies only on ability to *predict* whether a review will be helpful. In this experiment a random forrest (RF) classifier, based on all features described in Table 1, was used. As a simple baseline recommendation approach, a review was also selected at random (referred to as the *Rand* strategy).

The performance of these recommendation strategies can be evaluated in two ways. First, since the actual helpfulness scores of all reviews (the ground-truth) is known, the recommended review according to each strategy can be compared to the review which has the highest actual helpfulness score for each product, and averaged across all products in a given product category. Thus, the two line graphs in Fig. 6 plot the actual helpfulness of the recommended reviews (for *Pred* and *Rand*) as a percentage of the actual helpfulness of the most helpful review for each product; this is referred to as the *helpfulness ratio (HR)*. It can be seen that *Pred* significantly outperforms *Rand* delivering a helpfulness ratio of 0.9 and above compared to approximately 0.7 for *Rand*. This means that the *Pred* strategy is capable of recommending a review that has, on average, a helpfulness score which is 90% that of the actual most helpful review.

Incidentally, very often the most helpful review has a perfect helpfulness score of 1.0 and this review is often recommended by *Pred*. In this regard, the recommendation performance of the *Pred* and *Rand* strategies can be further analysed by examining how often, on average, each strategy recommends a review for each product from among the top k reviews ranked by actual helpfulness. In Fig. 6, results for $k = 3$ are presented (as bars) for each product class. For instance, it can be seen that for Laptops *Pred* recommends a top-3 review 60% of the time compared to only 37% for *Rand*. Moreover, across all product categories, the *Pred* strategy recommends a top-3 review between 1.5 and 2 times as frequently as *Rand*.

In summary, the above findings indicate that the helpfulness classifier can be used to recommend helpful reviews, without the need for explicit helpfulness information, and that recommendation performance compares favourably to the optimal scenario in which recommendations are based on known helpfulness information. These

Fig. 6 The average helpfulness ratio and top-k results for *Pred* and *Rand* across all product categories

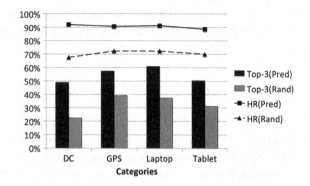

findings bode well for systems where review helpfulness is not available or is incomplete: it may still be possible to identify and recommend those reviews (new or old) which are likely to be genuinely helpful to users.

6 Case-Study 3: Mining Experiences and Recommending Products

Thus far, the focus of this chapter has been on user-generated reviews: their opinions, classification, and recommendation. In this case-study, however, the focus is changed from the reviews to the products being reviewed. After all, reviews exist because they reflect the experiences of users with real products and they are made available to users to help them chose a product for purchase. It makes sense therefore to consider whether the type of information mined from reviews, as described previously, can be aggregated at the level of individual products and used during classical product recommendation.

For instance, at the time of writing the listing for a *13" Retina MacBook Pro* on Amazon.com included a range technical features such as *screen-size, RAM, processor speed*, and *price*. These are the type of features that one might expect to find in a conventional content-based recommender system [29]. But in many domains such features are difficult to locate or are highly technical in nature, thereby limiting recommendation opportunities or making it difficult for casual consumers to judge the relevance of suggestions. However, the *MacBook Pro* has more than 70 reviews which encode valuable insights into a great many of its features, many of which are far from technical; for example, its *"beautiful design"*, its *"great video editing"* capabilities, and its *"high price"*. These features capture more detail than a handful of technical (catalog) features and in this case-study these *experiential* features (and associated sentiment) are used to build alternative product descriptions for use in a product recommender; this case-study is based on a series of research papers and further detail can be found in [30–33].

6.1 From Reviews Topics to Product Features

The reviews for each product, P, are converted into a rich, feature-based description (or *product case*) using the techniques described in Sect. 3: unigram and bi-gram features are extracted from each product review and sentiment scores are assigned to these features.

Thus, for each product P we now have a set of features $F(P) = \{F_1, \ldots, F_m\}$ extracted from the reviews of P ($Reviews(P)$), and how frequently each feature F_i is associated with positive, negative, or neutral sentiment in the particular reviews in $Reviews(P)$ that discuss F_i. For the purpose of this work features which are

mentioned in $\geq 10\%$ of reviews for that product are only considered and overall
sentiment (Eq. 5) and popularity (Eq. 6) scores are calculated; $Pos(F_i, P)$ (resp.
$Neg(F_i, P)$, $Neut(F_i, P)$) denotes the number of times that feature F_i has positive
(resp. negative, neutral) sentiment in the reviews for product P. The product case,
$Case(P)$, is then given by Eq. 7.

$$Sent(F_i, P) = \frac{Pos(F_i, P) - Neg(F_i, P)}{Pos(F_i, P) + Neg(F_i, P) + Neut(F_i, P)} \quad (5)$$

$$Pop(F_i, P) = \frac{|\{R_k \in Reviews(P) : F_i \in R_k\}|}{|Reviews(P)|} \quad (6)$$

$$Case(P) = \{[F_i, Sent(F_i, P), Pop(F_i, P)] : F_i \in F(P)\} \quad (7)$$

6.2 Recommending Products

We will consider a *more-like-this* product recommendation setting in which the user
is considering a particular product, Q, which serves as a *query product* for the
purpose of recommendations, generating a set of suggestions for similar products.
The above product representation leads to a content-based recommendation approach
based on feature similarity to the query product. However, the availability of feature
sentiment suggests another approach in which products that offer *better* quality
features compared to the query product can be recommended.

6.2.1 Similarity-Based Recommendation

Each product case is represented as a vector of features, where feature *values* represent their popularity in reviews (Eq. 6) as a proxy for their importance. The cosine
similarity between query product, Q, and candidate recommendation, C, is given by:

$$Sim(Q, C) = \frac{\sum_{F_i \in F(Q) \cup F(C)} Pop(F_i, Q) \times Pop(F_i, C)}{\sqrt{\sum_{F_j \in F(Q)} Pop(F_j, Q)^2} \sqrt{\sum_{F_j \in F(C)} Pop(F_j, C)^2}} \quad (8)$$

Using this approach, a set of top n recommendations are generated, ranked according to similarity with the query product [29].

6.2.2 Sentiment-Enhanced Recommendation

Rather than recommend products using *similarity* alone, feature sentiment can also be used to seek products with *better* sentiment than the query product. Equation 9 computes a score for feature F_i between query product Q and recommendation candidate C; a positive (resp. negative) score means that C has higher (resp. lower) sentiment for F_i compared to Q.

$$better(F_i, Q, C) = \frac{Sent(F_i, C) - Sent(F_i, Q)}{2} \tag{9}$$

Equation 10 computes an average better score at the product level across the *shared* features between Q and C. However, this approach ignores any *residual features* that are unique to Q or C. Thus, Eq. 11 computes an average better score across the *union* of features in Q and C; non-shared features are assigned a neutral sentiment score of 0.

$$B1(Q, C) = \frac{\sum_{F_i \in F(Q) \cap F(C)} better(F_i, Q, C)}{|F(Q) \cap F(C)|} \tag{10}$$

$$B2(Q, C) = \frac{\sum_{F_i \in F(Q) \cup F(C)} better(F_i, Q, C)}{|F(Q) \cup F(C)|} \tag{11}$$

6.2.3 Combining Similarity and Sentiment

The sentiment-based approaches above prioritise products that enjoy more positive reviews across a range of features relative to the query product. However, these recommendations may not necessarily be very similar to the query product. Thus, Eq. 12 ranks recommendations based on their combined (controlled by w) similarity and sentiment with respect to Q; $Bx(Q, C)$ denotes $B1(Q, C)$ or $B2(Q, C)$, normalised to $[0, 1]$.

$$Score(Q, C) = (1 - w)\, Sim(Q, C) + w \left(\frac{Bx(Q, C) + 1}{2} \right) \tag{12}$$

6.3 Evaluation

The above approaches are evaluated using data extracted from Amazon.com during October 2012. We considered 6 product domains in total but here present representative results for 3 domains (Table 3). For each product with ≥ 10 reviews, we extracted review texts, helpfulness information, and the top n ($n = 5$) recommendations for 'related' products as suggested by Amazon. In this case, related products

Table 3 Dataset statistics

Domain	#Reviews	#Products	#Features $\mu\,(\sigma)$	Sims $\mu\,(\sigma)$
Tablets	17,936	166	26 (10)	0.6 (0.1)
Phones	14,860	257	9 (5)	0.5 (0.2)
GPS	12,115	119	24 (11)	0.6 (0.2)

are those as suggested by Amazon's "customers who viewed this item also viewed these items" approach to recommendation.

6.3.1 Methodology and Metrics

A standard *leave-one-out* approach is used in our evaluation, comparing our recommendations for each product to those produced by Amazon. Thus, for each product (referred to as the query product, Q) in a given domain, a set of top-5 recommendations is generated using Eq. 12, varying w from 0 to 1 in steps of 0.1. This produces 22 recommendation lists for each Q, 11 each for $B1$ and $B2$, which are compared to Amazon's recommendations for Q.

Amazon's overall product ratings are used as an independent measure of product quality. The *ratings benefit* metric compares two sets of recommendations based on their ratings (Eq. 13), where a ratings benefit of 0.1 means that sentiment-based recommendations (R) enjoy an average rating score that is 10% higher that those produced by Amazon (A).

$$Ratings\ Benefit(R, A) = \frac{\overline{Rating(R)} - \overline{Rating(A)}}{\overline{Rating(A)}} \tag{13}$$

The *query product similarity* is also computed, given by the average similarity (by Eq. 8) based on mined feature representations between recommendations and the query product. This allows us to evaluate whether the sentiment-based techniques produce recommendations that are related to the query product and also provides a basis for comparison to Amazon's recommendations.

6.3.2 Mining Rich Product Descriptions

The success of our approach depends on its ability to translate user-generated reviews into useful product cases. Table 3 also shows the mean and standard deviation of the number of features that are extracted for each domain. On average, 9-26 features are extracted per product case, indicating that reasonably feature-rich cases are generated. Table 3 (last column) also shows the mean and standard deviation of the pairwise product cosine similarities. Again the results bode well because they show a relatively wide range of similarity values; very narrow ranges would suggest limitations in the expressiveness of extracted product representations.

6.3.3 Sentiment Versus Similarity

For each domain, Fig. 7a–c shows $B1$ and $B2$ results for top 5 recommendations. Ratings benefit scores (left y-axis, dashed lines) for $B1$ (circles) and $B2$ (squares) against w (x-axis), along with the corresponding query product similarity values (right y-axis, solid lines). The average similarity between the query product and the Amazon recommendations is also shown, which is independent of w and so appears as a solid horizontal line in each graph.

At $w = 0$, Eq. 12 is equivalent to a pure similarity-based approach to recommendation (i.e. using cosine by Eq. 8), because sentiment is not contributing to the overall recommendation score. For this configuration there is little or no ratings benefit; the recommendations produced have very similar average ratings to those produced by Amazon. However, the recommendations that are produced are more similar to the query product, at least in terms of the features mentioned in reviews, than Amazon's own recommendations. For example, in the Phones domain (Fig. 7b) at $w = 0$, recommendations based on cosine have a query product similarity of 0.8 compared to 0.6 for Amazon's recommendations.

At $w = 1$, where recommendations are based solely on sentiment, a range of maximum positive ratings benefits (from 0.18 to 0.23) can be seen across all 3 product domains. $B2$ outperforms $B1$, except for GPS, indicating that the sentiment associated with residual (non-shared) features is important, at least for two of the three domains considered. Consider again the Phones domain (Fig. 7b) at $w = 1$, where ratings benefits of 0.11 and 0.21 are achieved for $B1$ and $B2$, respectively. Thus, products recommended by $B2$ enjoy ratings that are 21 % higher than Amazon's recommendations, an increase of almost one point on average for Amazon's 5-point scale.

However, these ratings benefits are offset by a drop in query product similarity. At $w = 1$, query product similarity falls below that of the Amazon recommendations. Thus, a tradeoff exists between ratings benefits and query product similarity.

6.3.4 Balancing Similarity and Sentiment

The relative contribution of similarity and sentiment is governed by w (Eq. 12). As w increases a gradual increase in ratings benefit for $B1$ and $B2$ is seen, especially at larger w, with $B2$ outperforming $B1$ except for GPS. The slope of the ratings benefit curves and the maximum benefit achieved is influenced by the ratings distribution in each domain. For example, Phones and Tablets have ratings distributions with relatively low means and high standard deviations. Thus, more opportunities for

Fig. 7 Ratings benefit (*left* y-axis and *dashed* lines) and query similarity (*right* y-axis and *solid* lines) versus w (x-axis) for the Laptops (**a**), Phones (**b**) and GPS (**c**) domains. $B1$ and $B2$ are presented as *circles* and *squares* on the line graphs respectively and the Amazon query similarity is shown as a *solid horizontal line*

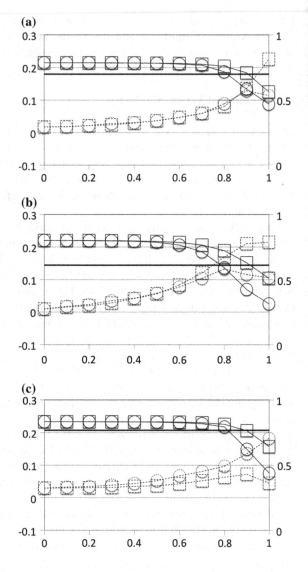

improved ratings exist and, indeed, the highest ratings benefits are seen for these domains (above 0.2 at $w = 1$ for $B2$).

Regarding query product similarity, there is little change for $w < 0.7$. But for $w > 0.7$ there is a reduction as sentiment tends to dominate during recommendation ranking. This query product similarity profile is remarkably consistent across all product domains and in all cases $B2$ better preserves query product similarity compared to $B1$.

To better understand the relative performance of $B1$ and $B2$ with respect to the Amazon baseline as w varies, a reference point is needed for the purpose of a

Fig. 8 Ratings benefits at Amazon baseline query product similarity

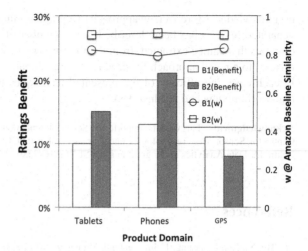

like-for-like comparison. To do this, we compare our techniques by fixing w at the point at which the query product similarity curve intersects with the Amazon query product similarity level and then reading the corresponding ratings benefits for $B1$ and $B2$. This is a useful reference point because it allows us to look at the ratings benefit offered by $B1$ and $B2$ when delivering recommendations that have the same query product similarity as the baseline Amazon recommendations.

Figure 8 shows these ratings benefits and corresponding w values for $B1$ and $B2$. The results clarify the positive ratings benefits that are achieved using sentiment-based recommendation without compromising query product similarity. For Tablets and Phones there are very significant ratings benefits, especially for $B2$ (resp. 15% and 21%). As stated above, $B1$ outperforms $B2$ for GPS, but in a relatively minor way, suggesting that the sentiment associated with residual features is not playing a significant role in this domain.

Finally, note the consistency of the w values at which the query product similarity of the sentiment-based recommendations matches that of Amazon. For each domain, $w \approx 0.9$ (for $B2$) delivers recommendations that balance query product similarity with significant ratings benefits; whether this value of w generalises to other domains is left to future work.

7 Conclusions

The web is awash with user-generated reviews, from the contemplative literary critiques of GoodReads to the flame wars that can sometimes engulf hotels on TripAdvisor. Reviews help consumers to choose and help online stores to convert browsers into buyers. In this chapter, a number of case-studies have been presented that focus on different ways to extract and harness the opinions contained in this valuable source of user knowledge. Moreover, an approach to opinion mining that is well suited to user-generated reviews has been described, and a number of useful applications for the opinions that can be extracted, from the filtering and recommendation of individual reviews to a novel approach for product recommendation, have been demonstrated. In each case, the efficacy of the presented techniques have been evaluated using real-world review and product data.

Acknowledgments This work is supported by Science Foundation Ireland: through the CLARITY Centre for Sensor Web Technologies under grant number 07/CE/I1147; and through the Insight Centre for Data Analytics under grant number SFI/12/RC/2289.

References

1. Hu, N., Liu, L., Zhang, J.: Inf. Technol. Manag. **9**, 201 (2008)
2. Zhu, F., Zhang, X.M.: J. Market. **74**(2), 133 (2010)
3. Kim, S.M., Pantel, P., Chklovski, T., Pennacchiotti, M. In: Proceedings of the Conference on Empirical Methods in Natural Language Processing (EMNLP 2006), pp. 423–430. Sydney, Australia (2006)
4. Liu, Y., Huang, X., An, A., Yu, X. In: Proceedings of the 8th IEEE International Conference on Data Mining (ICDM 2008), pp. 443–452. IEEE Computer Society, Pisa, Italy (2008)
5. Baccianella, S., Esuli, A., Sebastiani, F. In: Advances in Information Retrieval, 31th European Conference on Information Retrieval Research (ECIR 2009), pp. 461–472. Springer, Toulouse, France (2009)
6. Hsu, C.F., Khabiri, E., Caverlee, J. In: Proceedings of the 2009 IEEE International Conference on Social Computing (SocialCom 2009), pp. 90–97. Vancouver, Canada (2009)
7. O'Mahony, M.P., Cunningham, P., Smyth, B. In: Proceedings of the 20th Irish Conference on Artificial Intelligence and Cognitive Science (AICS 2009), pp. 244–253. Dublin, Ireland (2009)
8. O'Mahony, M.P., Smyth, B. In: Proceedings of the 3rd ACM Conference on Recommender Systems, RecSys '09. New York (2009)
9. Lim, E.P., Nguyen, V.A., Jindal, N., Liu, B., Lauw, H.W. In: Proceedings of the 19th ACM International Conference on Information and Knowledge Management, CIKM 2010, pp. 939–948. ACM, New York (2010). doi:10.1145/1871437.1871557. http://doi.acm.org/10.1145/1871437.1871557
10. Li, F., Huang, M., Yang, Y., Zhu, X. In: Proceedings of the 22nd International Joint Conference on Artificial Intelligence—Volume Volume Three, IJCAI 2011, pp. 2488–2493. AAAI Press, San Jose (2011). doi:10.5591/978-1-57735-516-8/IJCAI11-414. http://dx.doi.org/10.5591/978-1-57735-516-8/IJCAI11-414
11. Chevalier, J.A., Dina Mayzlin, D.: J. Market. Res. **43**(3), 345 (2006)
12. Dhar, V., Chang, E.A.: J. Interact. Market. **23**(4), 300 (2009)
13. Dellarocas, C., Zhang, M., Awad, N.F.: J. Interact. Market. **21**(4), 23 (2007)
14. O'Callaghan, D., Harrigan, M., Carthy, J., Cunningham, P. In: ICWSM (2012)
15. Popescu, A.M., Etzioni, O. In: Natural Language Processing and Text Mining, pp. 9–28. Springer, London, 2007. doi:10.1007/978-1-84628-754-12. http://dx.doi.org/10.1007/978-1-84628-754-1_2
16. Hu, M., Liu, B. In: Proceedings of the 10th ACM SIGKDD International Conference on Knowledge Discovery and Data Mining, KDD '04, pp. 168–177. ACM, New York (2004). doi:10.1145/1014052.1014073. http://doi.acm.org/10.1145/1014052.1014073
17. Hu, M., Liu, B. In: Proceedings of the 19th National Conference on Artifical Intelligence, AAAI'04, pp. 755–760. AAAI Press, San Jose (2004). http://dl.acm.org/citation.cfm?id=1597148.1597269

18. Zhang, K., Narayanan, R., Choudhary, A. In: Proceedings of the 3rd Workshop on Online Social Networks, WOSN '10. Berkeley (2010). http://dl.acm.org/citation.cfm?id=1863190.1863201
19. Garcia Esparza, S., O'Mahony, M.P., Smyth, B. In: Proceedings of the 4th ACM Conference on Recommender Systems, RecSys '10, pp. 305–308. ACM, New York (2010). doi:10.1145/1864708.1864773. http://doi.acm.org/10.1145/1864708.1864773
20. De Francisci Morales, G., Gionis, A., Lucchese, C. In: Proceedings of the 5th ACM International Conference on Web Search and Data Mining, WSDM '12, pp. 153–162. ACM, New York (2012). doi:10.1145/2124295.2124315. http://doi.acm.org/10.1145/2124295.2124315
21. Poirier, D., Tellier, I., Fessant, F., Schluth, J. In: Adaptivity, Personalization and Fusion of Heterogeneous Information, RIAO '10, pp. 136–137. Paris, France (2010). doi:http://dl.acm.org/citation.cfm?id=1937055.1937089
22. Justeson, J.S., Katz, S.M.: Nat. Lang. Eng. 1(1), 9 (1995)
23. Qiu, G., Liu, B., Bu, J., Chen, C. In: Proceedings of the 21st International Joint Conference on Artifical Intelligence, IJCAI 2009, pp. 1199–1204. Morgan Kaufmann Publishers Inc., San Francisco (2009). http://dl.acm.org/citation.cfm?id=1661445.1661637
24. Moghaddam, S., Ester, M. In: Proceedings of the 19th ACM International Conference on Information and Knowledge Management, CIKM '10, pp. 1825–1828. ACM, New York (2010). doi:10.1145/1871437.1871739. http://doi.acm.org/10.1145/1871437.1871739
25. DuBay, W. Impact Information, pp. 1–76 (2004)
26. O'Mahony, M.P., Smyth, B. In: Adaptivity, Personalization and Fusion of Heterogeneous Information, RIAO 2010, pp. 164–167. Paris, France (2010). http://dl.acm.org/citation.cfm?id=1937055.1937097
27. Witten, I., Frank, E. Data Mining: Practical Machine Learning Tools and Techniques. Morgan Kaufmann, San Diego (2005)
28. Streiner, D., Cairney, J. The Canadian Journal of Psychiatry/La revue Canadienne de Psychiatrie (2007)
29. Pazzani, M., Billsus, D. In: The Adaptive Web, Lecture Notes in Computer Science, vol. 4321, pp. 325–341. Springer, Berlin, Heidelberg (2007). doi:10.1007/978-3-540-72079-9_10. http://dx.doi.org/10.1007/978-3-540-72079-9_10
30. Dong, R., Schaal, M., O'Mahony, M.P., McCarthy, K., Smyth, B. In: Proceedings of the 20th International Conference on Case-Based Reasoning (2012), ICCBR '12, pp. 62–76
31. Dong, R., Schaal, M., O'Mahony, M.P., McCarthy, K., Smyth, B. In: Proceedings of the 21st International Conference on Case-Based Reasoning, ICCBR '13, vol. 7969, pp. 44–58. Springer, Heidelberg (2013)
32. Dong, R., Schaal, M., O'Mahony, M.P., Smyth, B. In: Proceedings of the 23rd International Joint Conference on Artificial Intelligence, IJCAI '13, pp. 1310–1316. AAAI Press, Menlo Park, California (2013)
33. Dong, R., O'Mahony, M.P., Schaal, M., McCarthy, K., Smyth, B. In: Proceedings of the 7th ACM Conference on Recommender Systems, RecSys '13, pp. 411–414. ACM, New York (2013). doi:10.1145/2507157.2507199. http://doi.acm.org/10.1145/2507157.2507199

Predicting Emotion Labels for Chinese Microblog Texts

Zheng Yuan and Matthew Purver

Abstract We describe an experiment into detecting emotions in texts on the Chinese microblog service Sina Weibo (www.weibo.com) using distant supervision via various author-supplied emotion labels (emoticons and smilies). Existing word segmentation tools proved unreliable; better accuracy was achieved using character-based features. Higher-order n-grams proved to be useful features. Accuracy varied according to label and emotion: while smilies are used more often, emoticons are more reliable. Happiness is the most accurately predicted emotion, with accuracies around 90 % on both distant and gold-standard labels. This approach works well and achieves high accuracies for happiness and anger, while it is less effective for sadness, surprise, disgust and fear, which are also difficult for human annotators to detect.

1 Introduction

Social media has become a very popular communication tool among Internet users. In China, the number of users of social networking websites had reached 288 million by the end of June 2013. The proportion of social networking service (SNS) users amongst Internet users was 48.8 % [5]. Sina Weibo (hereafter Weibo), is a Chinese microblog website. Most people take it as the Chinese version of Twitter; it is one of the most popular sites in China, with 60.2 million daily active users [6], and has therefore become a valuable source of people's opinions and sentiments.

Microblog texts (called *statuses* in Weibo) are very different from general newspaper or web text. Weibo statuses are shorter and more casual; many topics are

Z. Yuan (✉)
Computer Laboratory, University of Cambridge, Cambridge, UK
e-mail: Zheng.Yuan@cl.cam.ac.uk

M. Purver
School of Electronic Engineering and Computer Science,
Queen Mary University of London, London, UK
e-mail: m.purver@qmul.ac.uk

© Springer International Publishing Switzerland 2015
M.M. Gaber et al. (eds.), *Advances in Social Media Analysis*,
Studies in Computational Intelligence 602,
DOI 10.1007/978-3-319-18458-6_7

129

discussed, with less coherence between texts. Combining this with the huge amount of lexical and syntactic variety (misspelt words, new words, emoticons, unconventional sentence structures) in Weibo data, many existing methods for emotion and sentiment detection which depend on grammar- or lexicon-based information are no longer suitable.

Machine learning via supervised classification, on the other hand, is robust to such variety but usually requires hand-labelled training data. The labelling process is difficult and time-consuming with large datasets, and can be unreliable when attempting to infer an author's emotional state from short texts [31]. Our solution is to use *distant supervision*: we adapt the approach of [17, 31] to Weibo data, using emoticons and Weibo's built-in smilies as author-generated emotion labels for training, allowing us to learn a model of the associated language which can classify Weibo statuses into different basic emotion classes. Adapting this approach to Chinese data poses several research problems: finding accurate and reliable labels to use, segmenting Chinese text and extracting sensible lexical features.

Our experiments show that choice of labels has a significant effect, with emoticons generally providing higher accuracy than Weibo's smilies, and that choice of text segmentation method is crucial, with current word segmentation tools providing poor accuracy on microblog text and character-based features proving superior.

2 Background

2.1 Sentiment Analysis and Emotion Detection

Most research in this area focuses on sentiment analysis—classifying text as positive or negative [27]. However, finer-grained emotion detection is required to provide cues for further human-computer interaction, and is critical for the development of intelligent interfaces. It is hard to reach a consensus on how the basic emotions should be categorised, but here we follow [8] and others in using the definition in the work of [11], providing six basic emotions: anger, disgust, fear, happiness, sadness, and surprise.

Algorithms previously used for this task range from matching words in a sentiment lexicon to training classifiers with labelled data. In early work, Turney [41] used mutual information between document phrases and the word "excellent" and "poor" to get the average sentiment orientation of reviews. They used unsupervised classification and achieved an average accuracy of 74 %. Phrases containing adjectives or adverbs were extracted and used since they are good indicators of subjective [19]. Pang et al. [28] first applied different machine learning methods to detect the polarity of movie reviews. They reported the effectiveness of using machine learning techniques for sentiment classification: machine learning approach beats human-produced baselines easily. However, the performance was not as good as traditional topic-based text classification. They evaluated three machine learning methods (Naïve Bayes (NB), Maximum Entropy (ME) and Support Vector Machines

(SVMs)) and results showed that unigram presence information seemed to be the most effective. Yessenov and Misailovic [45] used movie review comments from social network Digg,[1] and evaluated both supervised learning (NB, ME, Decision trees) and unsupervised learning (K-Means). In addition to a bag-of-words model, they also tried to incorporate WordNet synonyms information. They came to a similar conclusion with [28] that the simple bag-of-words model performs relatively well. Tsutsumi et al. [40] proposed a way of using a multiple classifier based on three different classifiers. Results showed that the integrated methods outperformed all three single classifiers.

2.2 Distant Supervision

Distant supervision is an approach which combines standard supervised classification methods with a weakly labelled training dataset; it can be seen as an example of semi-supervised learning in that it exploits large amounts of data without access to expert gold-standard labels. Go et al. [17] and Pak and Paroubek [26], following [32], use emoticons in Twitter messages to provide these weak (or *noisy*) labels, then learn a classifier on the basis of the remaining text (after removal of the emoticons) to classify positive/negative sentiment with above 80 % accuracy.

Yuasa et al. [46] showed that emoticons have an important role in emphasizing the emotions conveyed in a sentence; they can therefore give us direct access to authors' own emotions. Derks et al. [10] and Provine et al. [29] similarly found that emoticons tend to increase the intensity of the associated verbal content, rather than replacing it (perhaps playing a similar role to laughter, facial expressions and other non-verbal behaviour). We would therefore expect them to be suitable for use as labels in a distant supervision approach, indexing the emotional content while leaving its verbal expression largely unaffected when the emoticons are removed. Purver and Battersby [31] investigated the applicability of this approach to English Twitter messages, using a broader set of emoticons to extend the distant supervision approach to six-way emotion classification, and we apply a similar approach here to Chinese Weibo statuses. However, in addition to the widely used, domain-independent emoticons, other markers have emerged for particular interfaces or domains. Weibo provides a built-in set of smilies that can work as special emoticons that help us better understand authors' emotions.

2.3 Chinese Text Processing

In Chinese text, sentences are represented as strings of Chinese characters without explicit word delimiters as used in English (e.g., white space). Therefore, it is

---

[1] http://digg.com.

important to determine word boundaries before running any word-based linguistic processing on Chinese.

There is a large body of research into Chinese word segmentation [12, 15, 18, 21, 35, 43]. These methods can be roughly classified into two categories: lexicon-based method and character-tagging method.

The idea for lexicon-based method is "segmentation". The basic technique for identifying distinct words is based on the lexicon-based identification scheme [4]. This approach performs the word segmentation process by using matching algorithms: matching input character strings with a known lexicon. However, since the real-world lexicon is open-ended, new words are coming out every day—and this is especially true with social media. A lexicon is therefore difficult to construct or maintain accurately for such a domain.

The character-tagging method was first introduced by [44]. It is more like a "word-building" process: it treats the word segmentation as a sequence labeling problem by assigning labels to all characters. Labels indicate whether a character locates at the beginning of, inside or at the end of a word. Several discriminative sequential learning algorithms have been exploited (e.g., conditional random fields (CRFs) [39], latent variable CRFs [37], structured perceptron [20], and the Passive-Aggressive algorithm [36]). However, the performance on social media data is not satisfying as the data is so different from the existing training libraries used.

3 Weibo Corpus

3.1 Corpus Collection

Our training data consisted of Weibo statuses with emoticons or smilies (see Sect. 3.2). Since Weibo has a public API,[2] training data can be collected through automated means. To use the API, we also need to create a Weibo account and register an application. We wrote a Python script which requested the *statuses public_timeline API*[3] every 30 s and inserted the collected data into a *MongoDB*[4] database. We constructed a corpus of Weibo data, filtering out messages not containing emotion labels (see Sects. 3.2 and 3.4 for details).

3.2 Emotion Labels

Two kinds of emotion labels (emoticons and smilies) were used as noisy labels. By "noisy", we mean that the emoticons and smilies are noisy themselves compared to

---

[2]http://open.weibo.com/wiki/API/en.
[3]http://open.weibo.com/wiki/2/statuses/public_timeline/en.
[4]http://www.mongodb.org/.

Fig. 1 Screenshot of the first page of Weibo built-in smilies

Table 1 Emoticons: Eastern style versus Western style

Emotion Classes	Eastern Style	Western Style
Happiness/Smile	(ˆ_ˆ)	:)
Sadness/Cry	(T_T)	:(
Anger	(ˇ_ˇ)	:@

gold-standard manual labels: to some degree ambiguous or vague in their meaning. Not all emoticons and smilies are closely related to these six emotion classes considered in our work; and some emoticons or smilies may be used differently in different situations, as people have different understandings. Smilies are Weibo built-in smilies (see Fig. 1) which form a finite, fixed set defined by the Weibo interface. Emoticons here are Eastern-style emoticons, which are made up of several characters and can thus be defined by the user; note that they are very different from Western-style emoticons [23] (see Table 1).

Eastern-style and Western-style emoticons are different, mostly because of different habits from using very different languages. For Western-style emoticons, people are used to reading them from left to right: Western emoticons are generally taken as being rotated by 90 degrees [30]. They are usually made of two to four characters and are of a relatively small number, generally focussing on some feature of mouth shape. Eastern emoticons, in contrast, are usually un-rotated and present faces, gestures, or postures from a point of view easily comprehensible to the reader.

At the beginning, we looked at all Eastern-style emoticons and Weibo built-in smilies available. Initial investigation found that not all emoticons and smilies can be classified into Ekman's six emotion classes [11]; and for some less frequently used labels, authors have widely different understandings. We therefore identified the most widely used and well-known emoticons/smilies; to then determine whether these would be reliable as labels, we set up a web survey to examine whether people could classify these emoticons/smilies consistently.[5]

Our survey contained two parts. In the first part, we asked people to choose one from the six emotion classes that best matched each of our identified emoticons/smilies. We also provided a *None of the above* option allowing participants to give their own definitions. In the second part, we asked people to tick all the emoticons and smilies they would use to convey each of the six emotions; we also allowed them to fill in other emoticons/smilies of their own that they would use for each emotion class. The survey was distributed via Weibo and only Chinese Weibo users were allowed to take part. 56 individuals completed our survey in two days time and full results are given in Appendix Table 9.

From the results of this, we identified 12 emoticons and 10 smilies to use as emotion labels (see Table 2). It is worth noting that we found no reliable emoticons for disgust, nor any reliable labels of either kind for fear. One reason may be that both disgust and fear, as emotion classes, are themselves difficult to represent (as facial expressions) using only punctuation and letters. For fear, we even found no relevant smilies in the Weibo interface. We believe this is because there is no obvious distinguishing feature on a fear face. In addition, people seem to use other emotions with fear, like "nervous", "cry". In order to ensure a reliable labelling, we decided to use only one smiley for disgust, and the keyword 害怕 for fear (a Chinese word meaning fear). However, we should be careful with keywords as they might not work well. Removing a word from a text may affect the meaning of the message itself and leave the rest of the text less informative and reliable. In addition, words are verbal, so they are subject to things like negation. Using keywords as emotion labels may be less reliable and it may result in lots of false positive examples.

3.3 Text Processing

Initial investigation also found that some Weibo statuses are mixtures of different language units: as well as Chinese, English words were also sometimes present and provided useful infomation. Therefore, in our work, not only Chinese characters/words, but also any lexical items from other languages were included as features. Weibo usernames (starting with @) and URLs were removed. Punctuation was included as a feature (treated like a lexical unigram), with any repeated punctuations being normalised to 3 characters. We then removed the labelling emoticons and smilies from the texts, using them instead only as positive/negative labels for the relevant emotion classes for training and testing purposes. We then extracted different kinds

---

[5]Available at: http://www.sojump.com/jq/1935017.aspx?npb=1.

Table 2 Conventional markers used for emotion classes

Emotion Classes	Emoticons	Smilies
Anger	(˘︹)	😠 [怒 nù "Anger"]
		😡 [怒骂 nù mà "Curse"]
Disgust	N/A	😷 [吐 tù "Spit"]
Fear	N/A	N/A
Happiness	(*^‿^*)	😊 [嘻嘻 xī xī "Hee hee"]
	(*^_^*)	😃 [哈哈 hā hā "Haha"]
	(*^o^*)	👏 [鼓掌 gǔ zhǎng "Applaud"]
	o(n_n)o	😄 [大开心 dà kāi xīn "So happy"]
	o(^_^)o	
	(^o^)	
	(^_^)	
Sadness	(T_T)	😭 [泪 lèi "Tear"]
	(T.T)	😢 [悲伤 bēi shāng "Sad"]
	(π.π)	
Surprise	(OMG)	😲 [吃惊 chī jīng "Surprise"]

of lexical features: segmented Chinese words, Chinese characters, and higher-order n-grams.

To use word-based features, we need to segment the statuses into words. There are lots of Chinese word segmentation tools; however, many are unsuitable for online social media text; we compared *Pymmseg*,[6] *Smallseg*[7] and *Stanford Chinese Word Segmenter*,[8] which all appeared to give reasonable results. *Pymmseg* uses the MMSEG algorithm [38]. *Smallseg* is an open sourced Chinese segmentation tool based on DFA. *Stanford Segmenter* is CRF-based [39].

3.4 Corpus Analysis

Our corpus contains 1,027,853 Weibo statuses with emotion labels; Table 3 shows statistics. The number of Weibo statuses varied with the popularity of labels themselves: labels for happiness and sadness are much more frequent than others;

---

[6]https://code.google.com/p/pymmseg-cpp/.

[7]https://code.google.com/p/smallseg/.

[8]https://nlp.stanford.edu/software/segmenter.shtml.

Table 3 Number of Weibo statuses per emotion class

Emotion classes	Using emoticons only	Using smilies only	Using both labels
Anger	427	60,271	60,698
Disgust	0	8,463	8,463
Fear	Using keyword 39,978[a]		
Happiness	19,979	529,077	549,056
Sadness	38,676	307,427	346,103
Surprise	3,097	20,458	23,555

[a]For "fear", we used the Chinese keyword 害怕 as the emotion label—see Sect. 3.2

very similar results were observed on English Twitter (see e.g., [31]), suggesting that these frequencies are relatively stable across very different languages.

Overall frequencies show that users of Weibo are more likely to use built-in smilies rather than emoticons. One possible reason is that smilies can be inserted with a single mouse click, whereas emoticons must be typed using several keystrokes—Eastern-style emoticons are usually made of five or more characters.

4 Experiments and Discussions

Machine learning techniques have been shown to be effective for traditional text classification and sentiment analysis. Here, we use Support Vector Machines (SVMs) [42], a state-of-the-art supervised kernel method. The basic idea is to find a maximum-margin hyperplane—a hyperplane that can separate two different classes correctly, and simultaneously maximize the margin (or the distance) between that hyperplane and other "difficult points" close to the hyperplane. These "difficult points" are called support vectors, and the decision function is fully specified by these support vectors. New testing examples are then assigned to one side of the hyperplane. Classifiers trained using SVMs have been shown to have better performance than other classifiers: Joachims [22] proved that SVMs consistently achieved good performance on text categorization tasks and outperformed other methods substantially and significantly; Pang et al. [28] applied different machine learning methods to detect the polarity of movie reviews. By evaluating three machine learning methods: Naïve Bayes (NB), Maximum Entropy (ME) and SVMs, they showed that SVMs had the best performance and NB turned out to be the worst. SVMs are good for high-dimensional feature spaces [22], while, other classifiers are training expensive when dealing with a large number of features.

In our work, classification was using SVMs throughout, with the help of LIB-LINEAR [13]. LIBLINEAR inherits many features of LIBSVM [3], but is more efficient for training large-scale problems without using kernels. The performance was evaluated using 10-fold cross validation.

Cross validation is used to estimate how well a model generalises [24]. For one round of cross validation, the dataset is partitioned into two subsets, one for training (*training set*) and one for testing (*validation set* or *testing set*). Several rounds of cross validation are performed, with different partitionings, in order to assess variance. Then we average the results and calculate the standard deviation (σ). F-fold cross validation was introduced by [16]. A single dataset is divided into F chunks; in each fold, 1 chunk is retained as the validation data (*test set*) while the remaining ($F - 1$) chunks are used as training data (*training set*). This process is repeated F times so that each of the F chunks is used exactly once as a test set.

Our training datasets were balanced: a dataset of size N contained $N/2$ positive instances (Weibo statuses containing labels for this emotion class) and $N/2$ negative ones (Weibo statuses containing labels from other classes). For $N/2$ negative instances, we randomly selected instances from other emotion classes for larger datasets ($N > 50,000$), but ensured an even weighting across negative classes for smaller sets to prevent bias towards one negative class.

4.1 Feature Selection

An important part of data-driven approach is converting a piece of text (the "observation") into a feature vector for text processing. A suitable feature vector should be designed and it should contain as few features as necessary. There is lots of work addressing the feature extraction problem for machine learning (e.g., see [14, 33]). In this section, we focused on two types of lexical features: word-based features and character-based features.

4.1.1 Word-Based Features

Chinese is written without spaces between words. In order to identify lexical features, we need to segment them first. Classification performance depends largely on the quality of the lexical features we obtain from different Chinese word segmentation tools.

However, people might find it difficult to apply existing segmentation tools to social media data. On one hand, unconventional words are used in microblogs: misspelt words, cyber words, as well as new words (see e.g., [1]). On the other hand, there are some pre-defined structures which are not used in other domains: Weibo usernames (@username), hashtags (#topic#), URLs, emoticons, smilies, etc.

For these latter unconventional (but known) structures, we can treat them separately, removing them before passing through the segmenter. However, for unconventional and misspelled words, this is not possible in general, and it is difficult for existing tools to identify them correctly. It may require better segmentation algorithms and new models should be trained using social media data. We investigated the effect of three different segmentation tools and results are presented in Fig. 2.

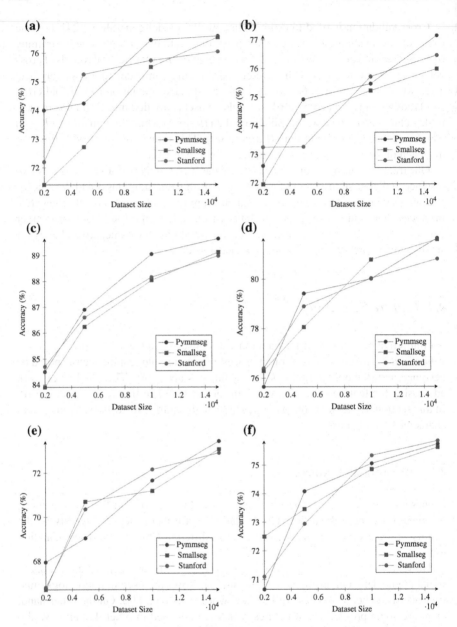

Fig. 2 Classification results of word-based features based on different segmentation tools. **a** Anger, **b** disgust, **c** fear, **d** happiness, **e** sadness, **f** surprise

Results showed that *Pymmseg* outperformed *Smallseg* and *Stanford Segmenter* for all emotion classes except `surprise` (where *Stanford Segmenter* yielded the best performance) as training dataset size increased. We can also learn from the results that accuracy increased as we using more training examples (see Sect. 4.2). We also want to point out that in terms of segmentation speed, *Pymmseg* is the fastest and *Stanford Segmenter* is the slowest. Therefore, we used *Pymmseg* for later experiments.

4.1.2 Character-Based Features

For character-based features, rather than requiring word segmentation, we simply treat each Chinese character as a unigram feature, as well as each punctuation character, emoticon and smiley (see Table 4).

Whether higher-order n-grams are useful features appears to be a matter of some debate. Pang et al. [28] reported that unigrams outperformed bigrams when classifying movie reviews by sentiment polarity, but [9] found that bigrams and trigrams can give better product-review polarity classification.

In our experiments with higher-order n-grams, we also included lower-order n-grams (e.g., for 5-grams, we used all unigrams, bigrams, trigrams, 4-grams and 5-grams as features, see Table 4), as there are lots of Chinese words with only one character.

Table 4 An example of one Weibo status and its n-gram features: repeated punctuations were normalised to 3 chars and reserved as a unigram; smiley was reserved as a unigram; Weibo username was removed

Weibo:	好饿！！！！！想吃东西的举手 [泪] @飞飞飞鸟_sunshinebird
unigram:	好 饿 !!! 想 吃 东 西 的 举 手 [泪]
bigram:	好 饿 !!! 想 吃 东 西 的 举 手 [泪] 好饿 饿!!! !!!想 想吃 吃东 东西 西的 的举 举手 手[泪]
trigram:	好 饿 !!! 想 吃 东 西 的 举 手 [泪] 好饿 饿!!! !!!想 想吃 吃东 东西 西的 的举 举手 手[泪] 好饿!!! 饿!!!想 !!!想吃 想吃东 吃东西 东西的 西的举 的举手 举手[泪]
4-gram:	好 饿 !!! 想 吃 东 西 的 举 手 [泪] 好饿 饿!!! !!!想 想吃 吃东 东西 西的 的举 举手 手[泪] 好饿!!! 饿!!!想 !!!想吃 想吃东 吃东西 东西的 西的举 的举手 举手[泪] 好饿!!!想 饿!!!想吃 !!!想吃东 想吃东西 吃东西的 东西的举 西的举手 的举手[泪]
5-gram:	好 饿 !!! 想 吃 东 西 的 举 手 [泪] 好饿 饿!!! !!!想 想吃 吃东 东西 西的 的举 举手 手[泪] 好饿!!! 饿!!!想 !!!想吃 想吃东 吃东西 东西的 西的举 的举手 举手[泪] 好饿!!!想 饿!!!想吃 !!!想吃东 想吃东西 吃东西的 东西的举 西的举手 的举手[泪] 好饿!!!想吃 饿!!!想吃东 !!!想吃东西 想吃东西的 吃东西的举 东西的举手 西的举手[泪]

For higher-order n-grams, lower-order n-gram features were also included

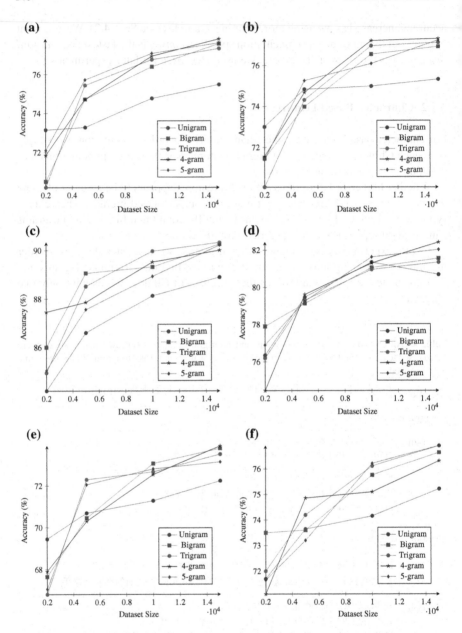

Fig. 3 Classification results of character-based n-gram features. **a** Anger, **b** disgust, **c** fear, **d** happiness, **e** sadness, **f** surprise

Results showed that higher-order n-grams are useful features for our wide-topic social media Weibo data. Higher-order n-grams (bigrams, trigrams, 4-grams and 5-grams) outperformed unigrams for all emotion classes by a large margin (see Fig. 3).

We stopped at 5-gram since the accuracy didn't improve any more. And as we adding higher-order n-gram features, it took more time to train classifiers.

4.1.3 Word-Based Features Versus Character-Based Features

Looking at all six emotion classes, we found that word-based features did not beat character-based ones. Character-based higher-order n-gram features had better performance than word-based features (even using the most effective segmenter, *Pymmseg*) for all emotion classes except `sadness`—see Table 5.

Our results suggested that we could just use Chinese characters, rather than doing any word segmentation. Three out of six emotion classes achieved their best performance by using character-based 4-gram features: `disgust`, `fear`, and `happiness`.

Examination of the segmented data showed that these three segmentation tools didn't work well with our social media data and made lots of segmentation mistakes. In addition, they produced many segmented words which contained only one character. The use of character-based features was therefore preferred and 4-gram features were used in later experiments.

4.2 Increasing Dataset Size

So far, experiments results also showed that increasing dataset sizes increased accuracy up to $N = 15,000$ (see Figs. 2 and 3). In this experiment, we kept increasing

Table 5 Classification accuracy for all six emotion classes ($N = 15,000$). The best one for each emotion class is marked in **bold**

No. of features[a]		Accuracy (%)					
		Anger	Disgust	Fear	Happiness	Sadness	Surprise
Word-based (Pymmseg)							
Unigram	45,103	76.59	77.14	89.65	81.65	73.45	75.74
Bigram	161,816	77.31	77.39	89.95	82.04	74.23	76.10
Trigram	261,070	77.21	76.62	90.07	81.79	**74.29**	76.43
4-gram	331,667	77.01	76.91	90.47	82.20	73.75	76.68
5-gram	394,352	77.49	77.47	90.17	81.97	73.27	76.00
Character-based							
Unigram	12,983	75.90	75.27	88.31	80.53	72.10	75.73
Bigram	139,897	77.17	77.33	90.27	81.77	73.17	75.92
Trigram	339,969	77.06	76.77	90.21	82.23	73.51	**77.05**
4-gram	498,838	77.29	**77.83**	**90.56**	**82.36**	73.73	75.75
5-gram	616,744	**77.89**	77.62	90.31	82.08	74.12	76.39

[a]For higner-order n-grams ($n > 1$), we removed features below a certain frequency threshold ($f = 2$)

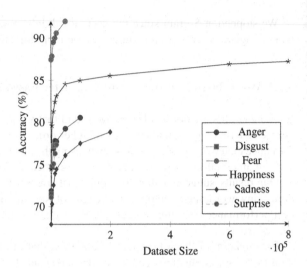

Fig. 4 Classification results for all six emotion classes

training dataset sizes for all six emotion classes and compared their classification results. Character-based 4-gram features were used, and as mentioned before, for larger datasets ($N > 50,000$), we randomly selected negative training examples from other emotion classes (see Sect. 4).

Because of the unbalanced number of Weibo statuses for each emotion class (see Sect. 3.4), the largest training dataset size for each emotion class varied: from $N = 15,000$ for disgust to $N = 800,000$ for happiness. Classification accuracy (using cross-validation) increased as we added more training examples, and does not appear to approach an asymptote until the largest sizes—see Fig. 4 and Table 6. As our dataset sizes increase over time, we therefore expect improvements in accuracy for all six emotion classes.

However, performances are quite different (see Table 6): fear is the most accurately predicted emotion (92.01 %) with the keyword as emotion label, followed by happiness (87.17 %), anger (80.56 %), sadness (78.85 %), surprise (77.36 %) and disgust (77.31 %).

4.3 Emotion Labels

In all experiments above, we used a random sample of instances "labelled" with either emoticons or smilies. In this experiment, we compared these two different types of emotion labels (emoticons and smilies) in terms of their classification accuracy. Four kinds of training dataset were constructed and tested for happiness, sadness and surprise:

Table 6 Classification results (accuracy (%)) for all six emotion classes. The best one for each emotion class is marked in **bold**

Training sizes	Anger	Disgust	Fear	Happiness	Sadness	Surprise
2,000	71.85	71.55	87.45	74.45	67.90	71.10
5,000	74.72	74.68	87.86	79.62	70.30	74.86
10,000	76.93	77.21	89.52	81.30	72.54	75.10
15,000	77.81	**77.31**	90.01	82.43	73.91	76.31
20,000	77.75		90.60	83.12	74.44	**77.36**
50,000	79.25		**92.01**	84.55	76.09	
100,000	**80.56**			84.98	77.51	
200,000				85.54	**78.85**	
600,000				86.88		
800,000				**87.17**		

- A dataset only contained instances collected with emoticons;
- A dataset only contained instances collected with smilies;
- Half of the training examples were collected with emoticons and the other half were collected with smilies;
- The training examples were randomly selected from all the instances collected with both emoticons and smilies.[9]

Comparing the accuracies between these sets tells us which of the label types is used in a more consistent way: association with a more consistent distribution of words/characters will result in higher classification accuracy (accuracy of prediction of emotion label). Results (see Fig. 5) showed that emoticon labels were easier to classify than smilies. By examining a sample of the data directly, we found that people use emoticons in a more systematic or consistent way. They tend to use emoticons to tell others what their real emotions are (happiness, sadness etc.); on the other hand, they use smilies for a much bigger range of things, such as jokes, sarcasm, etc. Some people use smilies just to make their Weibo statuses more interesting and lively, apparently without any subjective feelings.

4.4 Manual Labelling

So far, we used only the distant ("noisy") labels for both training and testing. In other words, classification accuracy is strictly only a measure of ability to predict the noisy label's presence (i.e., use of an emoticon or smiley), rather than necessarily measuring the ability to predict the author's emotion. To examine how well the two correspond, we must test against human judgements.

---

[9]That is how we constructed our training datasets for previous experiments.

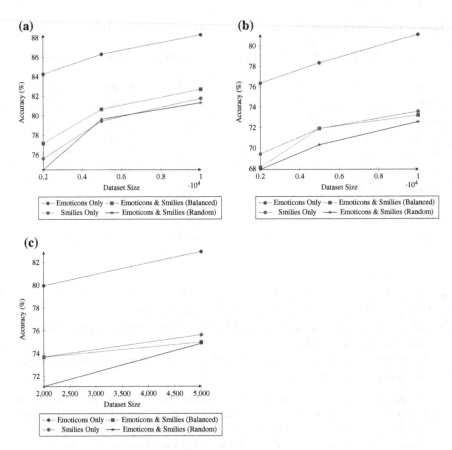

Fig. 5 Comparison of two different types of labels. Character-base 4-gram features were used. Performance was evaluated using 10-fold cross validation. **a** Happiness, **b** sadness, **c** surprise

Amazon's Mechanical Turk (MTurk)[10] service has shown to be useful for gathering human judgements for many simple NLP tasks (e.g., see [2, 7, 25, 34]). In our final experiment, we used MTurk to collect some manually labelled test data.

Another set of 2,190 instances was used for human annotation. These instances were collected using either emoticons or smilies, and were evenly distributed across our 6 emotion classes. Human annotators were asked to choose the strongest emotion class behind the message, with only one class allowed, although a *None of the above* option was also provided. Each instance was labelled by three different annotators.

Agreement between annotators was poor: only 26 % instances (571 out of 2,190) were assigned the same labels by all three annotators. These unanimous instances were quite unbalanced: from 5 examples for `fear` to 289 examples for `happiness`. When looking at instances agreed by a majority (i.e., at least two annotators), we got

---

1,335 (out of 2,190) examples varying from 27 for `fear` to 553 for `happiness`—see Table 7.

Two rounds of evaluation were performed where instances agreed by all and majority were used respectively. The best classifier for each emotion class from Sect. 4.2 was used. Since the test dataset was unbalanced, precision, recall and F1 for the class in question were used instead of accuracy. Recall is much higher than precision for some emotions (`sadness`, `surprise`, `disgust` and `fear`) when using default settings. In order to have a consistent F-score to compare between emotion classes, we also tuned these experiments so that recall approximate equals precision. Overall performance is shown in Table 8.

As before, results for `happiness` and `anger` are quite good, which showed that:

1. These two emotion classes are easier to detect;
2. The distant labels used for these two emotion classes are reliable;
3. Our classifiers are able to detect these two emotions.

Results for `surprise`, `sadness` and `disgust` can perhaps be considered reasonable, considering there are far fewer positive examples than negative ones in their test sets.

However, the result for `fear` is poor. Considering the low number of annotated positive test examples (see Table 7), we may conclude that this emotion class is

Table 7 Number of agreed instances for each emotion class

	Anger	Disgust	Fear	Happiness	Sadness	Surprise	All
Test 1[a]	93	26	5	289	103	55	571
Test 2[b]	216	102	27	553	267	170	1,335

[a]Labels of instances were agreed by all three annotators
[b]Labels of instances were agreed by at least two annotators

Table 8 Classification results on manually labelled data

(a) Test on instances agreed by all three annotators

	Anger (%)	Disgust (%)	Fear (%)	Happiness (%)	Sadness (%)	Surprise (%)
Precision	90.22	74.07	5.26	94.74	71.15	81.82
Recall	89.25	76.92	20.00	93.43	71.84	81.82
F1	89.73	75.47	8.33	94.08	71.50	81.82

(b) Test on instances agreed by at least two annotators

	Anger (%)	Disgust (%)	Fear (%)	Happiness (%)	Sadness (%)	Surprise (%)
Precision	72.43	61.39	48.15	87.70	64.71	69.46
Recall	71.76	60.78	48.15	88.97	65.92	68.24
F1	72.09	61.08	48.15	88.33	65.31	68.84

difficult to identify even for human annotators. It is interesting to note that our classifier failed to detect fear in these annotated examples even though it achieved high cross-validation accuracy (see Sects. 4.1 and 4.2). This was the only emotion category where we used the presence of a keyword, rather than a non-verbal sign (emoticon or smiley)—this suggests that the use of keywords is a poor method for distant supervision, as suspected.

5 Conclusion

In our work, we used SVMs for automatic emotion detection for Chinese microblog texts. We collected our own Weibo corpus and defined new emoticons and smilies as distant labels. Our results showed that using emoticons and smilies as noisy labels can be an effective way to perform distant supervision for Chinese, while the use of keywords extracted from the text is not effective. Emoticons seem to be more reliable for emotion detection than smilies.

It was also found that, when dealing with social media data, many existing Chinese word segmentation tools do not work well. Instead, we can use characters as lexical features and performance improves with higher-order n-grams. Character-based 4-gram features seem to be the most effective. Increasing the dataset size also improves performance, and our future work will examine larger sets.

Performance for different emotion classes are quite different: happiness is the most accurately predicted emotion (87.17 %), followed by anger (80.56 %). The effectiveness of our classifiers for these two emotion classes was also verified by using human annotated test data. Test results on manually labelled data also showed that the other four emotion classes (sadness, surprise, disgust and fear) are difficult to classify, either because reliable labels are hard to find (especially in the case of fear), and/or because they are difficult to detect even for human annotators.

Appendix

56 individuals completed our survey; the detailed results are presented here—see Table 9.

Table 9 Survey results showing the percentage of votes each emotion class received for each label. The best match for the defined labels used in our work are marked in **bold**

Emotion Labels	Anger	Disgust	Fear	Happiness	Sadness	Surprise	None
😠 [怒 nù "Anger"]	**85.71%**	1.79%	0	1.79%	0	0	10.71%
😡 [怒骂 nù mà "Curse"]	**73.21%**	3.57%	1.79%	1.79%	1.79%	0	17.86%
🤮 [吐 tù "Spit"]	1.79%	**58.93%**	0	0	1.79%	0	37.50%
😁 [嘻嘻 xī xī "Hee hee"]	0	0	0	**71.43%**	0	0	28.57%
😄 [哈哈 hā hā "Haha"]	0	0	0	**80.36%**	1.79%	0	17.86%
👏 [鼓掌 gǔ zhǎng "Applaud"]	0	0	1.79%	**73.21%**	0	0	25.00%
😃 [大开心 dà kāi xīn "So happy"]	0	0	1.79%	**73.21%**	0	0	25.00%
😢 [泪 lèi "Tear"]	1.79%	0	1.79%	0	**89.29%**	0	7.14%
😭 [悲伤 bēi shāng "Sad"]	0	1.79%	0	0	**89.29%**	0	8.93%
😲 [吃惊 chī jīng "Surprise"]	1.79%	0	3.57%	0	0	**76.79%**	17.86%
😤 [哼 hèng "humph"]	50.00%	19.64%	3.57%	1.79%	0	3.57%	21.43%
😒 [鄙视 bǐ shì "Despise"]	3.57%	35.71%	0	1.79%	0	0	58.93%
😞 [失望 shī wàng "Disappointed"]	0	1.79%	1.79%	0	53.57%	0	42.86%
(ˇ_ˇ)	**78.57%**	7.14%	0	0	1.79%	0	12.50%
(ˉ^ˉ)	39.29%	23.21%	0	0	14.29%	0	23.21%
('_')	1.79%	14.29%	10.71%	3.57%	33.93%	3.57%	32.14%
๙^๙	16.07%	3.57%	1.79%	3.57%	23.21%	0	51.79%
(*^_^*)	3.57%	0	0	**92.86%**	0	0	3.57%
(*ˉ_ˉ*)	1.79%	0	1.79%	85.71%	0	0	10.71%
(*ˉoˉ*)	0	0	1.79%	87.50%	0	1.79%	8.93%
o(n_n)o	0	0	0	89.29%	0	0	10.71%
o(ˉ_ˉ)o	3.57%	0	0	87.50%	0	0	8.93%
(ˇoˇ)	1.79%	1.79%	0	87.50%	1.79%	3.57%	3.57%
(ˇ_ˇ)	1.79%	1.79%	1.79%	89.29%	1.79%	0	3.57%
(T_T)	3.57%	0	1.79%	14.29%	67.86%	3.57%	8.93%
(T.T)	7.14%	3.57%	0	3.57%	60.71%	7.14%	17.86%
(TˉT)	14.29%	0	3.57%	12.50%	57.14%	0	12.50%
(π.π)	1.79%	3.57%	0	3.57%	80.36%	1.79%	8.93%
(OMG)	1.79%	0	0	1.79%	1.79%	**89.29%**	5.36%
(O_O)	0	1.79%	5.36%	10.71%	0	53.57%	28.57%
(O?O)	3.57%	0	1.79%	0	1.79%	39.29%	53.57%
(O.o)	3.57%	3.57%	0	1.79%	1.79%	57.14%	32.14%
(O_o)	1.79%	1.79%	3.57%	1.79%	1.79	55.36%	33.93%
(@_@)	0	0	3.57%	12.50%	1.79%	28.57%	53.57%
(*_*)	3.57%	1.79%	5.36%	16.07%	5.36%	14.29%	53.57%
(ಠ_ಠ)	19.64%	3.57%	7.14%	5.36%	8.93%	25.00%	30.36%
(ಠ_ಥ)	3.57%	8.93%	3.57%	5.36%	26.79%	16.07%	35.71%

References

1. Agichtein, E., Castillo, C., Donato, D., Gionis, A., Mishne, G. (2008) Finding high-quality content in social media. In: Proceedings of the 2008 International Conference on Web Search and Data Mining (WSDM'08). pp. 183–194
2. Bloodgood, M., Callison-Burch, C.: Bucking the trend: large-scale cost-focused active learning for statistical machine translation. In: Proceedings of the 48th Annual Meeting of the Association for Computational Linguistics, pp. 854–864. Uppsala, Sweden (2010)
3. Chang, C., Lin, C.: LIBSVM: a library for Support Vector Machines (2001). http://www.csie.ntu.edu.tw/cjlin/papers/libsvm.pdf. Cited 4 Feb 2014
4. Chen, K., Liu, S.: Word identification for Mandarin Chinese sentences. In: Proceedings of the 14th Conference on Computational Linguistics, (1992), vol. 1, pp. 101–107
5. China Internet Network Information Center (CINIC).: The 32nd Statistical Report on Internet Development in China (2013). http://www1.cnnic.cn/IDR/ReportDownloads/201310/P020131029430558704972.pdf. Cited 2 Feb 2014

6. China, SINA Corporation (SINA) Q3 2013 Earnings Conference Call (2013). http://seekingalpha.com/article/1835112-sina-corporations-ceo-discusses-q3-2013-results-earnings-call-transcript. Cited 2 Feb 2014
7. Callison-Burch, C.: Fast, Cheap, and Creative: evaluating translation quality using Amazons mechanical turk. In: Proceedings of the 2009 Conference on Empirical Methods in Natural Language Processing (EMNLP-2009), pp. 286–295. Singapore (2009)
8. Chuang, Z., Wu, C.: Multimodal emotion recognition from speech and text. Comput. Linguist. Chin. Lang. 9(2), 45–62 (2004)
9. Dave, K., Lawrence, S., Pennock, D.M.: Mining the peanut gallery: opinion extraction and semantic classification of product reviews. In: WWW2003, pp. 519–528
10. Derks, D., Bos, A., von Grumbkow, J.: Emoticons and online message interpretation. Soc. Sci. Comput. Rev. 26(3), 379–388 (2008)
11. Ekman, P.: Universal facial expressions of emotion. In: California Mental Health Research Digest, vol. 8, no. 4 (1970)
12. Fan, C., Tsai, W.: Automatic word identification in Chinese sentences by the relaxation technique. In: Computer Processing of Chinese and Oriental Languages (1988)
13. Fan, R., Chang, K., Hsieh, C., Wang, X., Lin, C.: LIBLINEAR: a library for large linear classification. J. Mach. Learn. Res. 9(2008), 1871–1874 (2008)
14. Forman, G.: An extensive empirical study of feature selection metrics for text classification. J. Mach. Learn. Res. 3, 1289–1305 (2003)
15. Gan, K., Palmer, M., Lua, K.: A statistically emergent approach for language processing: application to modeling context effects in ambiguous Chinese word boundary perception. Comput. Linguist. 22(4), 53153 (1996)
16. Geisser, S.: The predictive sample reuse method with applications. In: Journal of the American Statistical Association, pp. 320–328 (1975)
17. Go, A., Bhayani, R., Huang, L.: Twitter Sentiment Classification using Distant Supervision. Master's thesis, Stanford University (2009)
18. Guo, J.: Critical tokenization and its properties. Comput. Linguist. 23(4), 569596 (1997)
19. Hatzivassiloglou, V., Wiebe, J.M.: Effects of adjective orientation and gradability on sentence subjectivity. In: Proceedings of the 18th International Conference on Computational Linguistics (2000)
20. Jiang, W., Huang, L., Liu, Q.: Automatic adaptation of annotation standards: Chinese word segmentation and pos tagging a case study. In: Proceedings of the Joint Conference of the 47th Annual Meeting of the ACL and the 4th International Joint Conference on Natural Language Processing of the AFNLP, pp. 522–530. Suntec, Singapore (2009)
21. Jin, W., Chen, L.: Identifying unknown words in Chinese corpora. In: First Workshop on Chinese Language, University of Pennsylvania, Philadelphia (1998)
22. Joachims, T.: Text categorization with suport vector machines: learning with many relevant features. In: Proceedings of the 10th European Conference on Machine Learning (ECML'08), pp. 137–142 (1998)
23. Kayan, S., Fussell, S.R., Setlock, L.D.: Cultural differences in the use of instant messaging in Asia and North America. In: Proceedings of the 20th Anniversary Conference on Computer Supported Cooperative Work (CSCW'06), pp. 525–528. Banff, Alberta, Canada (2006)
24. Kohavi, R.: A study of cross-validation and bootstrap for accuracy estimation and model selection. In: Proceedings of the 14th International Joint Conference on Artificial Intelligence (IJCAI). Morgan Kaufmann, San Mateo (1995)
25. Nakov, P.: Noun compound interpretation using paraphrasing verbs: feasibility study. In: Proceedings of the 13th International Conference on Artificial Intelligence: Methodology, Systems and Applications (AIMSA 2008), pp. 103–117
26. Pak, A., Paroubek, P.: Twitter as a corpus for sentiment analysis and opinion mining. In: Proceedings of the 7th Conference on International Language Resources and Evaluation (LREC'10). Valletta, Malta (2010)
27. Pang, B., Lee, L.: Opinion mining and sentiment analysis. In: Foundations and Trends in Information Retrieval (2008)

28. Pang, B., Lee, L., Vaithyanathan, S.: Thumbs up? Sentiment classification using machine learning techniques. In: Proceedings of Empirical Methods in Natural Language Processing, (2002), pp. 79–86

29. Provine, R., Spencer, R., Mandell, D.: Emotional expression online: emoticons punctuate website text messages. J. Lang. Soc. Psychol. **26**(3), 299–307 (2007)

30. Ptaszynski, M., Maciejewski, J., Dybala, P., Rzepka, R., Araki, K.: CAO: A fully automatic emoticon analysis system based on theory of kinesics. In: Affective Computing, IEEE Transactions (2010)

31. Purver, M., Battersby, S.: Experimenting with distant supervision for emotion classification. In: Proceedings of the 13th Conference of the European Chapter of the Association for Computational Linguistics (EACL), pp. 482–491. Avignon, France (2012)

32. Read, J.: Using emoticons to reduce dependency in machine learning techniques for sentiment classification. In: Proceedings of the ACL Student Research Workshop, pp. 43–48. Ann Arbor, Michigan (2005)

33. Sebastiani, F.: Machine learning in automated text categorization. ACM Comput. Surv. **34**(1), 1–47 (2002)

34. Snow, R., O'Connor, B., Jurafsky, D., Ng, A.Y.: Cheap and fast but is it good? Evaluating non-expert annotations for natural language tasks. In: Proceedings of the 2008 Conference on Empirical Methods in Natural Language Processing (EMNLP-2008). Honolulu, Hawaii (2008)

35. Sproat, R., Shih, C.: A statistical method for finding word boundaries in Chinese text. In: Computer Processing of Chinese and Oriental Languages (1990)

36. Sun, W.: Word-based and characterbased word segmentation models: Comparison and combination. In: Coling 2010: Posters, pp. 1211–1219. Beijing, China (2010)

37. Sun, X., Zhang, Y., Matsuzaki, T., Tsuruoka, Y., Tsujii, J.: A discriminative latent variable Chinese segmenter with hybrid word/character information. In: Proceedings of Human Language Technologies: The 2009 Annual Conference of the North American Chapter of the Association for Computational Linguistics, pp. 56–64. Boulder, Colorado (2009)

38. Tsai, C.: MMSEG: A Word Identification System for Mandarin Chinese Text Based on Two Variants of the Maximum Matching Algorithm (2000). http://technology.chtsai.org/mmseg/. Cited 4 Feb 2014

39. Tseng, H., Chang, P., Andrew, G., Jurafsky, D., Manning, C.: A conditional random field word segmenter. In: Proceedings of the 4th SIGHAN Workshop on Chinese Language Processing (2005)

40. Tsutsumi, K., Shimada, K., Endo, T.: Movie review classification based on a multiple classifier. In: Proceedings of the 21st Pacific Asia Conference on Language, Information and Computation (PACLIC) (2007)

41. Turney, P.D.: Thumbs Up or Thumbs Down? Semantic orientation applied to unsupervised classification of reviews. In: Proceedings of the 40th Annual Meeting of the Association for Computational Linguistics (ACL), pp. 417–424. Philadelphia (2002)

42. Vapnik, V.N.: The Nature of Statistical Learning Theory (1995)

43. Wu, A.: Customizable segmentation of morphologically derived Words in Chinese. In: Computational Linguistics and Chinese Language (2003)

44. Xue, N.: Chinese word segmentation as character tagging. In: International Journal of Computational Linguistics and Chinese Language Processing (2003)

45. Yessenov, K., Misailovic, S.: Sentiment analysis of movie review comments. In: Methodology (2009), pp. 1–17

46. Yuasa, M., Saito, K., Mukawa, N.: Emoticons convey emotions without cognition of faces: an fMRI study. In: CHI 06 Extended Abstracts on Human Factors in ComputingSystems (2006), pp. 1565–1570

Author Index

© Springer International Publishing Switzerland 2015
M.M. Gaber et al. (eds.), *Advances in Social Media Analysis*,
Studies in Computational Intelligence 602,
DOI 10.1007/978-3-319-18458-6

Printed in the United States
By Bookmasters